Lecture Notes in Statistics

Edited by D. Brillinger, S. Fienberg, J. Gani,
J. Hartigan, and K. Krickeberg

12

W0051154

Martin Jacobsen

Statistical Analysis
of Counting
Processes

Springer-Verlag
New York Heidelberg Berlin

Martin Jacobsen
University of Copenhagen
Institute of Mathematical Statistics
5 Universitetsparken
DK-2100 Copenhagen Ø
Denmark

AMS Classification: 62L99, 62M99

Library of Congress Cataloging in Publication Data

Jacobsen, Martin, 1942-
    Statistical analysis of counting processes.

    (Lecture notes in statistics ; 12)
    Bibliography: p.
    Includes index.
    1. Stochastic processes. I. Title. II. Series:
Lecture notes in statistics (Springer-Verlag) ; v. 12.
QA274.J33 1982   519.5   82-19241

ISBN-13: 978-0-387-90769-7   e-ISBN-13: 978-1-4684-6275-3
DOI: 10.1007/978-1-4684-6275-3

Preface.

A first version of these lecture notes was prepared for a course given
in 1980 at the University of Copenhagen to a class of graduate students
in mathematical statistics. A thorough revision has led to the result
presented here.

The main topic of the notes is the theory of multiplicative intensity models for counting processes, first introduced by Odd Aalen in his
Ph.D. thesis from Berkeley 1975, and in a subsequent fundamental paper
in the Annals of Statistics 1978.

In Copenhagen the interest in statistics on counting processes was
sparked by a visit by Odd Aalen in 1976. At present the activities here
are centered around Niels Keiding and his group at the Statistical Research Unit.

The Aalen theory is a fine example of how advanced probability
theory may be used to develop a powerful, and for applications very relevant, statistical technique.

Aalen's work relies quite heavily on the 'theorie generale des
processus' developed primarily by the French school of probability theory. But the general theory aims at much more general and profound results, than what is required to deal with objects of such a relatively
simple structure as counting processes on the line. Since also this
process theory is virtually inaccessible to non-probabilists, it would
appear useful to have an account of what Aalen has done, that includes
exactly the amount of probability required to deal satisfactorily and
rigorously with statistical models for counting processes.

It has therefore been my aim to present a unified and essentially
selfcontained exposition of the probability theory for counting processes and its application to the statistical theory of multiplicative
intensity models. The inclusion of a purely probabilistic part conforms
with my view that to apply the Aalen models in practice, one must have
a thorough grasp of the underlying probability theory. Of course to

carry out this programme, some knowledge of probability must be pre-
supposed, especially conditional probabilities, weak convergence and
basic martingale theory.

The first three chapters deal with univariate and multivariate
counting processes and their probabilistic structure, while Chapters
4 and 5 are concerned with the definition of Aalen models and Aalen
estimators, and the asymptotic results required to make the models
applicable in statistical practice.

Naturally, the terminology and notation used in the general theory
of processes has been carried over to the special situation treated
here. One particularly relevant part of the general theory concerns
the definition and basic properties of stochastic integrals of predictable processes with respect to martingales. This in particular, is one
place where the setup involving only counting processes permits simpli-
fication compared to the general theory: whereas quite a lot of work
is required to define the general stochastic integrals, all the integrals appearing here are ordinary (random) Lebesgue-Stieltjes integrals.

A number of exercises are given at the end of each chapter. Some
of the exercises deal with proofs and arguments omitted from the text,
while others aim at covering part of the theory and examples not included elsewhere.

Notation. The notation $s\downarrow\downarrow t$ means that $s \to t$ with $s > t$, where
$s\downarrow t$ allows for $s \to t$ with $s \geq t$. For X a random variable defined
on a probability space $(\Omega, A, \mathbb{P})$, the notation $\mathbb{P}X$ rather than EX
is used for the expectation of X. Also, $\mathbb{P}(X;A)$ denotes the integral $\int_A X d\mathbb{P}$. Throughout $\mathbb{P}$ refers to a probability on some abstract probability space, while the letter P is reserved for probabilities on some specific spaces. The notation F, F_t for σ-algebras
and N_t for random variables also refers exclusively to these particular spaces.

Acknowledgements. I am especially indebted to Niels Keiding whose in-
formal notes for a course on counting processes he gave in 1977-78 have
been instrumental for the preparation of my own course in 1980, and
thereby also for the writing of these notes.

I would like to thank Per Kragh Andersen, Richard Gill, Inge Hen-
ningsen, Søren Johansen, Niels Keiding, Henrik Ramlau-Hansen as well as
my class, autumn 1980, for helpful discussions and comments.

The manuscript was typed at the Department of Mathematics, University of Copenhagen. I am most grateful for this essential assistance,
and I am happy to thank Dita Andersen and Jannie Larsen for the very
efficient job they have done.

Finally, thanks also to Simon Holmgaard for proofreading the entire
manuscript.

Copenhagen, March 1982

Martin Jacobsen

TABLE OF CONTENTS

1. ONE-DIMENSIONAL COUNTING PROCESSES   1

   1.1.   Probabilities on (0,∞]   1

   1.2.   The definition of one-dimensional counting processes   5

   1.3.   Construction of canonical counting processes   16

   1.4.   Intensities for canonical counting processes   26

   1.5.   Martingale decompositions for canonical counting   38
          processes

   1.6.   Statistical models and likelihood ratios   44

   Notes   47

   Exercises   48

2. MULTIVARIATE COUNTING PROCESSES   53

   2.1.   Definition and construction of multivariate counting 53
          processes

   2.2.   Intensities and martingale representations   63

   2.3.   Products of canonical counting processes   72

   2.4.   Likelihood ratios   74

   2.5.   Discrete counting processes   76

   Exercises   87

3. STOCHASTIC INTEGRALS   88

   3.1.   Processes and martingales on W^E 88

   3.2.   Definition and basic properties of stochastic   99
          integrals

   Notes   109

   Exercises   110

4. THE MULTIPLICATIVE INTENSITY MODEL   115

    4.1.   Definition of the full Aalen model   115

    4.2.   Product models and sufficient reductions   122

    4.3.   Estimation in the Aalen Model   128

    4.4.   Estimation in Markov chains   135

    4.5.   The Cox regression model   143

    4.6.   Maximum-likelihood estimation in Aalen models   148

    Notes   157

    Exercises   159

5. ASYMPTOTIC THEORY   161

    5.1.   A limit theorem for martingales   161

    5.2.   Asymptotic distributions of Aalen estimators   166

    5.3.   Asymptotic distributions of product-limit   181
            estimators

    5.4.   Comparison of two intensities   191

    Notes   195

    Exercises   198

APPENDIX   208

    1.   The principle of repeated conditioning   208

    2.   Weak convergence   212

REFERENCES   217

SUBJECT INDEX   223

1. ONE-DIMENSIONAL COUNTING PROCESSES

1.1. Probabilities on $(0,\infty]$.

Consider the half-line $(0,\infty]$ (0 excluded, ∞ included) equipped with the Borel σ-algebra B of subsets generated by the subintervals of $(0,\infty]$.

A probability Pr on $((0,\infty],B)$ may be described by its distribution function F, defined by $F(t) = Pr(0,t]$, $0 < t < \infty$. The function F is non-decreasing, right-continuous and satisfies $F \leq 1$, $\lim_{t\downarrow\downarrow 0} F(t) = 0$. If conversely F defined on $(0,\infty)$ is any function with these properties, then there is a unique probability Pr on $(0,\infty]$ with F as distribution function.

Instead of the distribution function F, one may use the survivor function $G = 1-F$. The following properties characterizes the survivor functions G for probabilities on $(0,\infty]$: G is non-increasing, right-continuous and satisfies $G \geq 0$, $\lim_{t\downarrow\downarrow 0} G(t) = 1$.

The termination point $t^\dagger$ of a probability on $(0,\infty]$ is defined as $t^\dagger = \inf\{t > 0: G(t) = 0\} = \sup\{t > 0: G(t) > 0\}$. Thus, if $t^\dagger < \infty$, $G(t^\dagger) = 0$ while $G(s) > 0$ for $s < t^\dagger$.

A probability on $(0,\infty]$ allows absorption if it has an atom at ∞: $Pr\{\infty\} > 0$. In that case $Pr\{\infty\} = G(\infty -) \overset{D}{=} \lim_{t\uparrow\uparrow\infty} G(t)$ is the absorption probability.

Suppose now that the probability Pr on $(0,\infty]$ is absolutely continuous (strictly speaking the restriction of Pr to $(0,\infty)$ is absolutely continuous with respect to Lebesgue measure) with density f, i.e. there is a non-negative, possibly infinite, measurable function f defined on $(0,\infty)$ such that

$$F(t) = \int_0^t ds\, f(s) \qquad (0 < t < \infty)$$

(equivalently, $G(t) = G(\infty -) + \int_t^\infty ds\, f(s)$ for $0 < t < \infty$).

We shall say that Pr has a <u>smooth density</u> if f: $(0,\infty) \to [0,\infty]$
may be chosen to be right-continuous with left-limits everywhere such
that $\lim_{t \downarrow \downarrow 0} f(t)$ exists (using the usual topology on $(0,\infty)$ and on
$[0,\infty]$ the topology obtained when adjoining ∞ to $[0,\infty)$ (with the
usual topology) in a one-point compactification).

Suppose Pr has a smooth density f . The <u>intensity</u> or <u>hazard</u>
for Pr is the function $\mu: (0,\infty) \to [0 \ \infty]$ defined by

$$\mu(t) = \begin{cases} f(t)/G(t) & \text{if } G(t) > 0 \\ 0 & \text{if } G(t) = 0 . \end{cases}$$

Since f is right-continuous one has, provided $G(t) > 0$, that

$$\mu(t) = \lim_{h \downarrow \downarrow 0} \frac{1}{h} Pr(t,t+h]/Pr(t,\infty]$$

so that, suitably normalized, the intensity $\mu(t)$ measures the risk
of "dying immediately after time t given survival up to t".

1.1. <u>Example.</u> Let $0 \leq \mu < \infty$ be a constant. The <u>exponential law</u>
with rate μ is the probability on $(0,\infty]$ with survivor function
$G(t) = \exp(-\mu t)$. It has smooth density $f(t) = \mu e^{-\mu t}$ and an intensity which is constant and equal to μ . The special case $\mu = 0$ corresponds to the probability degenerate at ∞ , (absorption probability 1). ∎

Expressed in terms of the survivor function G alone, it is seen
that

$$\mu = D^+(-\log G) ,$$

where D^+ is the right sided differential operator:
$D^+\phi(t) = \lim_{h \downarrow \downarrow 0} \frac{1}{h}(\phi(t+h) - \phi(t))$. Conversely G may be recovered from
μ by

(1.2) $$G(t) = \exp\left(-\int_0^t ds \ \mu(s)\right) \qquad (0 < t < \infty).$$

It should now be clear that the intensity function μ for probabilities on $(0,\infty]$ with smooth densities are characterized by the following properties: μ is non-negative, right-continuous everywhere with left-limits everywhere except possibly at $t^\dagger$, the limit $\lim_{t\downarrow 0}\mu(t)$ exists, μ is locally integrable at 0 in the sense that $\int_0^h ds\,\mu(s) < \infty$ for some $h > 0$, and finally $\mu(t) = 0$ whenever $\int_0^t ds\,\mu(s) = \infty$.

If $\Pr$ has intensity μ it is seen that 1): $\Pr$ has a finite termination point $t^\dagger$ iff μ is not locally integrable, i.e. $\int_0^t ds\,\mu(s) = \infty$ for some $0 < t < \infty$, and in that case $\int_0^{t^\dagger} ds\,\mu(s) = \infty$ and $\int_0^t ds\,\mu(s) < \infty$ for $t < t^\dagger$; 2): $\Pr$ has ∞ as termination point but does not allow absorption iff μ is locally but not globally integrable, i.e. $\int_0^t ds\,\mu(s) < \infty$ for $0 < t < \infty$ and $\int_0^\infty ds\,\mu(s) = \infty$; 3): $\Pr$ allows absorption iff μ is globally integrable, i.e. $\int_0^\infty ds\,\mu(s) < \infty$, and in that case the absorption probability equals $\exp\!\left(-\int_0^\infty ds\,\mu(s)\right)$.

If for some $t_0 > 0$, $\Pr(t_0,\infty] = 1$, then of course $\mu(t) = 0$ for $0 \le t < t_0$ and (1.2) may be written

$$G(t) = \exp\!\left(-\int_{t_0}^t ds\,\mu(s)\right) \qquad (t_0 \le t < \infty)$$

with $G(t) = 1$ for $t \le t_0$.

1.3. **Example.** If $\Pr$ has intensity μ, then for any $t_0 > 0$ the conditional probability $\Pr(\cdot \mid (t_0,\infty])$ has intensity function

$$\mu_{\mid t_0}(t) = \begin{cases} 0 & (0 < t < t_0) \\ \mu(t) & (t_0 \le t < \infty) \end{cases}$$

and survivor function

$$G_{\mid t_0}(t) = \begin{cases} 1 & (0 < t < t_0) \\ \exp(-\int_{t_0}^t ds\,\mu(s)) & (t_0 \le t < \infty). \end{cases} \qquad \Box$$

The following result will be useful later.

1.4. Proposition. Let T be a $(0,\infty]$-valued random variable such that the distribution of T has a smooth density with intensity μ and let $0 < \mu_0 < \infty$ be a constant. Then, assuming that $\int_0^\infty ds\, \mu(s) = \infty$, the random variable

$$U = \frac{1}{\mu_0} \int_0^T ds\, \mu(s)$$

follows an exponential law with rate μ_0.

Proof. Define $H(t) = \int_0^t ds\, \mu(s)$ and denote by H^{-1} the right-continuous inverse of H: $H^{-1}(u) = \inf\{t>0: H(t) > u\}$. Since $\int_0^\infty ds\, \mu(s) = \infty$, $H^{-1}(u)$ is defined for all $0 \le u < \infty$ and furthermore satisfies $H(H^{-1}(u)) = u$, $H(t) > u$ for $t > H^{-1}(u)$. Thus, if $\mathbb{P}$ denotes the probability on the probability space where T is defined, for any $0 \le u < \infty$

$$\mathbb{P}(U > u) = \mathbb{P}(H(T) > \mu_0 u) = \mathbb{P}(T > H^{-1}(\mu_0 u))$$

$$= \exp(-H(H^{-1}(\mu_0 u))) = e^{-\mu_0 u}. \qquad \blacksquare$$

1.2. The definition of one-dimensional counting processes.

A one-dimensional counting process may be thought of as a stochastic process recording at any given time t the number of certain
events having occurred before time t. This is formalized in Definition 2.1 below.

Let $(\Omega, A, A_t, \mathbb{P})$ be a <u>probability space with a filtration</u>, i.e.
$(\Omega, A, \mathbb{P})$ in a usual probability space and $(A_t)_{t \geq 0}$ is a family of
sub σ-algebras of A such that $A_s \subset A_t$ when $s \leq t$. A stochastic process $X = (X_t)_{t \geq 0}$ defined on (Ω, A) is <u>adapted</u> to (A_t) if
each X_t is A_t-measurable.

(Note: when writing $(I_t)_{t \geq 0}$ for some indexed family of objects,
the indexing set is $[0, \infty)$, so there is an I_t for each $0 \leq t < \infty$
but not apriori for $t = \infty$).

2.1. <u>Definition</u>. A <u>one-dimensional counting process</u> on a filtered
probability space $(\Omega, A, A_t, \mathbb{P})$, is an adapted stochastic process
$K = (K_t)_{t \geq 0}$, each K_t taking values in $\overline{\mathbb{N}}_0 = \{0, 1, \ldots, \infty\}$ with
$\mathbb{P}(K_0 = 0) = 1$ and such that almost all sample paths are non-decreas-
ing and right-continuous everywhere, increasing only by jumps of size
1.

The process is <u>stable</u> if $\mathbb{P}(K_t < \infty) = 1$ for all $t \geq 0$.
The process allows <u>absorption</u> if $\mathbb{P}(\sup_{t \geq 0} K_t < \infty) > 0$. |

Recall that the sample paths for $K = (K_t)_{t \geq 0}$ are the functions
$t \rightarrow K_t(\omega)$ obtained for any $\omega \in \Omega$. The definition demands that for
ω outside a $\mathbb{P}$-null set, the sample path determined by ω be right-
continuous. The topology on $\overline{\mathbb{N}}_0$ to be referred to when making this
statement precise is the one obtained by adjoining ∞ as the one-point
compactification to $\mathbb{N}_0 = \{0, 1, \ldots\}$, the set of non-negative integers,
equipped with the discrete topology.

It is readily checked that with this choice for the topology on $\mathbb{N}_0$, almost all sample paths will have left-limits everywhere.

Since we shall only discuss one-dimensional counting processes in this section we shall for simplicity refer to such a process as a counting process.

If we are just given a probality space $(\Omega, A, \mathbb{P})$ and a process $K = (K_t)_{t \geq 0}$ with almost all sample paths having the analytic proper-ties required by Definition 2.1, it is always possible to find a filtration $(K_t)_{t \geq 0}$ such that on $(\Omega, A, K_t, \mathbb{P})$ is a counting process: define $K_t = \sigma(K_s)_{s \leq t}$, the smallest sub σ-algebra of A with respect to which all K_s, $s \leq t$ become measurable. If we are given a count-ing process K on a filtered space $(\Omega, A, A_t, \mathbb{P})$, then K is also a counting process on $(\Omega, A, K_t, \mathbb{P})$ and $K_t \subset A_t$ so that $(K_t)_{t \geq 0}$ is the smallest filtration with respect to which K is a counting pro-cess. We shall call $(K_t)_{t \geq 0}$ the __self-exciting__ filtration for the process K.

Given a counting process $K = (K_t)_{t \geq 0}$, consider the mapping $\omega \to (K_t(\omega))_{t \geq 0}$ which to every $\omega \in \Omega$ associates the corresponding sample path of the process. This mapping T carries each ω into an element of the function space $\overline{\mathbb{N}}_0^{[0,\infty)}$ of all functions (paths) defined on $[0,\infty)$ taking values in $\overline{\mathbb{N}}_0$, which, for almost all ω, has specific analytic properties. Taking out a relevant subset W of $\mathbb{N}_0^{[0,\infty)}$ and equipping it as a measurable space one may therefore transform the original probability $\mathbb{P}$ on Ω into a probability $P = T(\mathbb{P})$ on W, which in a canonical fashion describes the probabi-litic properties of the process K. These considerations lead to Definitions 2.2 and 2.3.

__2.2. Definition.__ The __full counting process path-space__ is the subset $\overline{W}$ of $\overline{\mathbb{N}}_0^{[0,\infty)}$ consisting of those paths $w: [0,\infty) \to \overline{\mathbb{N}}_0$ with $w(0) = 0$ which are everywhere right-continuous and non-decreasing, increasing only in jumps of size 1.

The <u>stable counting process path-space</u> is the subset W of $\overline{W}$ consisting of those paths $w \in \overline{W}$ for which $w(t) < \infty$ for all $t \geq 0$.

From purely theoretical considerations, the full space $\overline{W}$ is the natural one to use as will be apparent from the next subsection. But for most statistical applications the stable space W is the appropriate one.

For $t \geq 0$, define $N_t: \overline{W}$ $(W) \to \overline{\mathbb{N}}_0$ by $N_t(w) = w(t)$ and let F denote the smallest σ-algebra of subsets of $\overline{W}$ (W) that makes all N_t measurable: $F = \sigma((N_t = n): n \in \mathbb{N}_0, t \geq 0)$. Also, let F_t be the σ-algebra generated by $(N_s)_{s \leq t}$. (Thus (F_t) is a filtration on $\overline{W}$ (W) and $N = (N_t)$ is adapted to this filtration). Note that $F_0 = \{\emptyset, \overline{W}\}$ $(\{\emptyset, W\})$. Finally define $N_\infty = \lim_{t \uparrow \uparrow \infty} N_t$.

It is possible to describe the σ-algebra F_t in a different way. On $\overline{W}$ (W) introduce an equivalence relation $\underset{t}{\sim}$ by requiring that $w \underset{t}{\sim} w'$ iff $w(s) = w'(s)$ for $0 \leq s \leq t$. Then $F \in F_t$ iff $F \in F$ and F is a union of $\underset{t}{\sim}$-equivalence classes, which are then also referred to as the <u>atoms</u> of F_t. (Sketch of proof: clearly the collection of sets which are F-measurable unions of $\underset{t}{\sim}$-equivalence classes, form a σ-algebra $\widetilde{F}_t$, and since obviously $(N_s = n) \in \widetilde{F}_t$ for $n \in \mathbb{N}_0$, $0 \leq s \leq t$, we have $F_t \subset \widetilde{F}_t$. Conversely the mapping $S_t: \overline{W} \to \overline{W}$ given by $N_s \circ S_t = N_{s \wedge t}$ $(s \geq 0)$ is F_t-measurable and has the property that $S_t^{-1} F = F$ for $F \in \widetilde{F}_t$ wherefore also $\widetilde{F}_t \subset F_t$). Using this equivalence class description of F_t, to show that a random variable defined on $(\overline{W}, F)$ $((W,F))$ is F_t-measurable, amounts to showing that it is constant on each F_t-atom. The σ-algebra F_t contains all information about the behaviour of the N_s on the time interval $[0,t]$. It is customary in general process theory to consider the slightly larger σ-algebras $F_{t+} \overset{D}{=} \bigcap_{\epsilon > 0} F_{t+\epsilon}$. However, in the case of the counting process path spaces $\overline{W}$ and $\overline{W}$, we have that $F_{t+} = F_t$, the reason being that knowing exactly the behaviour of a path w on $[0,t]$ tell us also the behaviour of w on $[0,t+\varepsilon]$ for some $\varepsilon > 0$, viz. $w(s) = w(t)$ for $t \leq s \leq t+\varepsilon$ by right-continuity. (Formally a proof that $F_{t+} = F_t$ may be given as follows: it is shown that F_{t+} consists of the sets which are F-measurable unions of equivalence classes for the equivalence relation $\underset{t+}{\sim}$ given by $w \underset{t+}{\sim} w'$ iff for some $\varepsilon = \varepsilon(w,w') > 0$, $w \underset{t+\varepsilon}{\sim} w'$; then it is observed that $\underset{t+}{\sim}$ is the same as $\underset{t}{\sim}$).

We have now equipped the path-spaces $\overline{W}$ and W with a measurable structure and are ready to give the next fundamental definition.

2.3. <u>Definition</u>. A <u>canonical one-dimensional counting process</u> is a probability on $(\overline{W}, F)$. A <u>stable canonical one-dimensional counting process</u> is a probability on (W, F). ∎

For convenience we shall abbreviate canonical counting process as CCP.

Thus, for CCP's the family of random variables defining the process is always the family (N_t) of projections and a CCP is cahracterized exclusively as a probability on $\overline{W}$ or W.

If P is a CCP we shall also use the symbol P to denote P-expectation. Thus, if $F \in F$ and U is real-valued and F-measurable we write $P(F)$, $P(U)$, $P(U;F)$ for respectively the P-measure of the set F, the integral $\int dP \, U$ and the integral $\int_F dP \, U$.

Note that any CCP, P, is completely determined by its collection of finite-dimensional distributions, i.e. the P-distribution of any vector $(N_{t_1}, \dots, N_{t_r})$ where $r \in \mathbb{N}$, $0 \leq t_1 < \dots < t_r$.

Suppose that $K = (K_t)_{t \geq 0}$ is a counting process on $(\Omega, A, A_t, \mathbb{P})$ in the sense of Definition 2.1. Taking away a $\mathbb{P}$-null set N, the mapping T discussed above becomes a measurable mapping from

$(\Omega \smallsetminus N, A(\Omega \smallsetminus N))$ to $(\overline{W}, F)$ (to (W, F) if K is stable) and hence induces a probability $P = T(\mathbb{P})$ on $(\overline{W}, F)$ $((W, F))$, the <u>canonical counting process generated</u> by K.

By the transformation some information may have been lost, but all information contained in the process itself has been retained: for every $t \geq 0$, knowing the restriction of P to F_t determines the restriction of $\mathbb{P}$ to K_t, and complete knowledge of P determines the restriction of $\mathbb{P}$ to K, the smallest sub σ-algebra of A containing the members K_t of the self-exciting filtration.

In these notes we shall mainly be concerned with CCP's. In statistical terms this means that we shall consider only the counting process itself as observable.

2.4. <u>Example</u>. The most important of all counting processes is the <u>Poisson process</u>. For $0 < \mu < \infty$ a constant, the canonical Poisson process with rate (or intensity) μ is the probability Π_μ on the stable space (W, F) with respect to which $(N_t)_{t \geq 0}$ has stationary independent Poisson increments: for $r \in \mathbb{N}$, $0 = t_0 < t_1 < \ldots < t_r$, $n_1, \ldots, n_r \in \mathbb{N}_0$

$$\Pi_\mu(N_{t_i} - N_{t_{i-1}} = n_i, \; i=1, \ldots, r) = \prod_{i=1}^{N} \Pi_\mu(N_{t_i} - N_{t_{i-1}} = n_i)$$

and for $0 \leq s < t$, $n \in \mathbb{N}_0$

$$\Pi_\mu(N_t - N_s = n) = \frac{(\mu(t-s))^n}{n!} e^{-\mu(t-s)} .$$

These distributional properties may also be written

$$\Pi_\mu(N_u - N_t = n | F_t) = \frac{(\mu(u-t))^n}{n!} e^{-\mu(u-t)}$$

for $0 \leq t \leq u$, $n \in \mathbb{N}_0$. |

A CCP P is <u>Markov</u> if for all $t < u$, $n \in \mathbb{N}_0$

$$P(N_u = n | F_t) = P(N_u = n | N_t),$$

i.e. if P makes (N_t) a Markov process. The transition probabilities for this Markov process are

$$p_{mn}(t,u) = P(N_u = n | N_t = m)$$

defined for all $m \le n \in \mathbb{N}_0$, $t \le u$ such that $P(N_t = m) > 0$. The process P is <u>stationary Markov</u> if P makes (N_t) Markov with stationary probabilities, i.e. if $p_{mn}(t,u)$ depends on t and u only through the difference $u-t$.

2.5. <u>Example</u>. The Poisson process Π_μ is stationary Markov with transition probabilities

$$p_{mn}(t,u) = \frac{(\mu(u-t))^{n-m}}{(n-m)!} e^{-\mu(u-t)} . \qquad \blacksquare$$

In Section 1.3 we shall briefly discuss (almost) all canonical Markov counting processes.

2.6. <u>Example</u>. Let $X_1,\ldots,X_r$ be random variables taking values in $(0,\infty)$. Thinking of the X_i as lifetimes of individuals, define for $t \ge 0$, K_t as the number of individuals that died at age t or earlier:

$$K_t = \sum_{i=1}^r 1_{(X_i \le t)} .$$

Provided no two individuals can die at the same time, $K = (K_t)$ is a counting process with respect to the self-exciting filtration. Notice that observing K alone amounts to observing the order statistics $(X_{(1)},\ldots,X_{(r)})$, but that there is no information in K about which individual died first, which died second etc. $\qquad \blacksquare$

2.7. Example. Suppose that in addition to the X_i we are given censoring random variables $U_1, \cdots, U_r$, taking values in $(0, \infty]$, such that for individual i the time of death is observed only if $X_i \leq U_i$. Then the number of individuals observed to die at age t or earlier is

$$K_t = \sum_{i=1}^{r} 1(X_i \leq t \wedge U_i),$$

and we get a counting process based on censored survival data.

Observing only the process K means that most of the original information about the times U_i of censoring will be lost, except of course if the U_i are assumed to take given fixed values. Phrased differently, the process K is adequate for studying censored survival times if one works conditionally given the values of the U_i. If that is not reasonable one may instead have to introduce a second counting process, recording the number of censorings taking place before time t. (See Example 2.1.6).

A classical problem in the statistical analysis of survival data consists in assuming the X_i to be independent and identically distributed, the $U_i = u_i$ to be given constants, and then estimating in a nonparametric fashion the common distribution function of the X_i. Without censoring (all $u_i = \infty$) the natural estimator is of course the empirical distribution function. With censoring, allowing the distribution of the X_i to be anything and then maximizing the probability of the actual observation, one arrives at the Kaplan-Meier estimator, (Kaplan and Meier (1958)).

The Kaplan-Meier estimator $\hat{G}$ of the survivor function for the X_i may conveniently be written

$$\hat{G}(t) = \prod_{0 < s \leq t} \left(1 - \frac{\Delta K_s}{R_{s-}}\right),$$

where $\Delta K_s = K_s - K_{s-}$ is the jump of K at s, and R_{s-} is the size of the population at risk at time s, i.e. the number of individuals observed to be alive (not censored and not dead) immediately before s . Formally thus

$$R_{s-} = \sum_{i=1}^{r} 1_{(s \leq X_i \wedge u_i)} ,$$

and here as elsewhere we use the notation R_{s-} rather than just R_s to stress that the process R_- is left-continuous.

Of course the continuous product defining $\hat{G}$ is just a finite product, the factors entering at the times, where the process K jumps.

We shall later in greater generality discuss the product-limit estimators of which $\hat{G}$ is an example. As for the interpretation of $\hat{G}$, suffice it for the moment to say that $\hat{G}$ is the survivor function for a discrete probability on $(0, \infty]$ with $\Delta K_s / R_{s-}$ the estimated discrete intensity or hazard at time s , i.e. the estimated probability of dying at s given survival up to s . ▌

We shall now in greater detail discuss the structure of the paths in the path-spaces $\overline{W}$ and W and also introduce some new concepts.

Suppose $w \in \overline{W}$. By assumption $w(0) = 0$ and by right-continuity $w(t) = 0$ for $t > 0$ sufficiently small. If w ever leaves the initial state 0 it happens with a jump to 1 where, again by right-continuity, the path must stay for a strictly positive amount of time, then jumping, if jumping at all, to 2 etc. Thus the path w may be completely described by a non-decreasing sequence $(\tau_n(w))_{n \geq 1}$ of strictly positive, possibly infinite, real numbers such that $\tau_n(w) < \tau_{n+1}(w)$ whenever $\tau_n(w) < \infty$ and where $\tau_n(w)$ denotes the timepoint of the n'th jump of w , defined to be ∞ if w has fewer than n jumps. If $\tau_\infty(w) \overset{D}{=} \lim_{n \uparrow \infty} \tau_n(w) < \infty$, then since w is non-decreasing, $w(t) = \infty$ for $t \geq \tau_\infty(w)$.

More formally, define $\tau_0 \equiv 0$ and for $n = 1, 2, \cdots$ define $\tau_n : \overline{W} \rightarrow (0, \infty]$ by

$$\tau_n = \inf\{t > 0 : N_t = n\}$$

with the convention $\inf \emptyset = \infty$. Also define $\tau_\infty = \lim_{n \uparrow \infty} \tau_n$.

Notice that as a subset of $\overline{W}$, $W = (\tau_\infty = \infty)$.

2.8. Definition. A random time is a measurable mapping $\tau: \overline{W} \to [0,\infty] (\tau: W \to [0,\infty])$.

A stopping time is a random time such that $(\tau = t) \in F_t$ for all $t \geq 0$.

The definition of a stopping time shows that it is possible to tell whether τ takes the value t_0, simply by observing the process (N_t) on the time interval $[0, t_0]$.

2.9. Proposition. A mapping $\tau: \overline{W} \to [0,\infty] (\tau: W \to [0,\infty])$ is a stopping time if and only if $(\tau \leq t) \in F_t$ for all $t \geq 0$ or if and only if $(\tau < t) \in F_t$ for all $t > 0$.

Proof. If $(\tau \leq t) \in F_t$, then $(\tau < t) = \bigcup_{n \geq 1} (\tau \leq t - \frac{1}{n}) \in F_t$ since $(\tau \leq t - \frac{1}{n}) \in F_{t - \frac{1}{n}} \subset F_t$. If $(\tau < t) \in F_t$ for all $t > 0$, then $(\tau \leq t) = \bigcap_{n \geq m} (\tau < t + \frac{1}{n})$ for any $m \geq 1$ and since $(\tau < t + \frac{1}{n}) \in F_{t + \frac{1}{n}} \subset F_{t + \frac{1}{m}}$ for $n \geq m$ we get $(\tau \leq t) \in F_{t + \frac{1}{m}}$ for every m and therefore $(\tau \leq t) \in F_{t+}$. But as noted earlier $F_{t+} = F_t$ and so $(\tau \leq t) \in F_t$.

Thus the two conditions $(\tau \leq t) \in F_t$ for $t \geq 0$ and $(\tau < t) \in F_t$ for $t > 0$ are equivalent and we need only show that the first is equivalent to saying that τ is a stopping time.

If $(\tau \leq t) \in F_t$ for $t \geq 0$, then τ is certainly a random time, and since $(\tau = t) = (\tau \leq t) \setminus \bigcup_{n=1}^{\infty} (\tau \leq t - \frac{1}{n})$, clearly $(\tau = t) \in F_t$ and τ is a stopping time.

If conversely τ is a stopping time, then certainly $(\tau \leq t) \in F$, so to show that $(\tau \leq t) \in F_t$, it suffices to show that $(\tau \leq t)$ is a union of F_t-atoms. But if $w \in (\tau \leq t)$,

1.2.10

$w' \underset{t}{\sim} w$, then $w' \underset{s}{\sim} w$ where $s = \tau(w) \le t$, so by the definition of a stopping time, $w' \in (\tau = s) \subset (\tau \le t)$ as desired. ∎

2.10. Proposition. All of $\tau_1, \tau_2, \cdots, \tau_\infty$ are stopping times.

Proof. Just observe that for $n \in \mathbb{N}$, $(\tau_n \le t) = (N_t \ge n)$ and that $(\tau_\infty \le t) = \underset{n \ge 1}{\cap} (\tau_n \le t)$. ∎

It is convenient to have a concept describing the information in a CCP up to a random time τ.

2.11. Definition. For τ a random time, the pre-τ algebra is the sub σ-algebra of F consisting of sets which are F-measurable unions of equivalence classes for the equivalence relation $\underset{\tau}{\sim}$ on $\overline{W}$ (W) given by $w \underset{\tau}{\sim} w'$ iff $\tau(w) = \tau(w')$ and $w(s) = w'(s)$ for $0 \le s \le \tau(w)$. ∎

Note that if $\tau(w) = \infty$, then the F_τ-atom containing w is simply the one-point set $\{w\}$.

2.12. Example. If τ is constant, equal to t where $0 \le t < \infty$, then $F_\tau = F_t$. ∎

2.13. Example. If $\tau = \tau_n$, then F_τ is the σ-algebra generated by $(\tau_1, \cdots, \tau_n)$. If $\tau = \tau_\infty$, then F_τ is the σ-algebra generated by $(\tau_1, \tau_2, \cdots)$ which is F itself. ∎

Usually, for τ a stopping time (but not otherwise), F_τ is defined by

$$F_\tau = \{F \in F: F(\tau \le t) \in F_t \text{ for all } t \ge 0\}.$$

The proof that this description agrees with Definition 2.11 is left as an exercise.

Given a random time τ, the position of the process (N_t) at that random time is defined by the mapping $N_\tau : \overline{W} \to \overline{\mathbb{N}_0}$ given by $N_\tau(w) = N_{\tau(w)}(w)$, (in particular $N_\tau = N_\infty$ on $(\tau = \infty)$).

2.14. Proposition. For τ a random time, N_τ is F_τ-measurable.

Proof. Obviously N_τ is constant on the F_τ-atoms, so we need only show that N_τ is F-measurable. But by the right-continuity of paths

$$N_\tau = \lim_{n \to \infty} \sum_{k=0}^{\infty} N_{\frac{k+1}{2^n}} 1_{\left(\frac{k}{2^n} \leq \tau < \frac{k+1}{2^n}\right)} + N_\infty 1_{(\tau = \infty)}$$

(interpreting $0 \cdot \infty$ as 0), and everything on the right-hand side is F-measurable. ∎

1.3. Construction of canonical counting processes.

Since a path $w \in \overline{W}$ is completely determined by the sequence $(\tau_n(w))_{n \geq 1}$ of jump times, a canonical counting process P is determined by the P-distribution of (τ_n). This again may be specified by giving the distribution of τ_1, the conditional distribution of τ_2 given τ_1, the conditional distribution of τ_3 given (τ_1, τ_2) etc. Obviously, given F_{τ_n} (i.e. given $(\tau_1, \ldots, \tau_n)$, see Example 2.13), the conditional distribution of τ_{n+1} is concentrated on the time interval $(\tau_n, \infty]$ provided we are inside the set $(\tau_n < \infty)$. As such the conditional distribution is given by a survivor function on $(\tau_n, \infty]$.

With this in mind the following basic construction result is not surprising.

3.1. __Theorem__. Suppose given for $n \in \mathbb{N}_0$ and any $0 < t_1 < \ldots < t_n < \infty$ a probability concentrated on the interval $(t_n, \infty]$ with survivor function $G_{nt_1 \ldots t_n}$ such that the collection of probabilities satisfy that for every $t > 0$ the mapping $(t_1, \ldots, t_n) \to G_{nt_1 \ldots t_n}(t)$ is measurable. Then there is a unique canonical counting process P such that for $n \in \mathbb{N}_0$, $t > 0$

$$(3.2) \qquad P(\tau_{n+1} > t | F_{\tau_n}) = G_{n\tau_1 \ldots \tau_n}(t)$$

P-a.s. on $(\tau_n < \infty)$.

__Proof__. Given the collection $G_{nt_1 \ldots t_n}$ of survivor functions, construct on a suitable probability space $(\Omega, A, \mathbb{P})$ a sequence of strictly positive, possibly infinite random variables $T_1, T_2, \ldots$ such that

$$\mathbb{P}(T_{n+1} > t | T_1, \ldots, T_n) = G_{nT_1 \ldots T_n}(t).$$

Because $\lim_{t \downarrow \downarrow t_n} G_{nt_1 \ldots t_n}(t) = 1$, this sequence will with probability one have the property that $T_n < \infty$ implies $T_{n+1} > T_n$. Let

$$\Omega_0 = \{\omega \in \Omega : T_{n+1}(\omega) > T_n(\omega) \quad \text{for all} \quad n \in \mathbb{N} \quad \text{with} \quad T_n(\omega) < \infty\}$$

so that $\mathbb{P}\Omega_0 = 1$. Now define a counting process $K = (K_t)_{t \geq 0}$ on $(\Omega_0, A\Omega_0, \mathbb{P})$ by $K_t = \sup\{n \in \mathbb{N}_0 : T_n \leq t\}$ where $T_0 \equiv 0$. Then the CCP generated by K satisfies (3.2). It is unique by the remarks preceding the theorem. ∎

Remark. For $n = 0$ there is just one survivor function G_0 in the collection $G_{nt_1 \ldots t_n}$ which specifies the marginal distribution of τ_1, so in this case (3.2) reads

$$P(\tau_1 > t) = P(\tau_1 > t | F_{\tau_0}) = G(t)$$

which is fine if one puts $\tau_0 \equiv 0$ and recalls that $F_0 = \{\emptyset, \overline{W}\}$. ∎

It should be emphasized that the theorem provides a construction of CCP's on $\overline{W}$, not on W. This means that the $G_{nt_1 \ldots t_n}$ may be chosen completely freely as survivor functions for probabilities on $(t_n, \infty]$. In Example 3.4 and in the next section we shall discuss various conditions for obtaining stable CCP's.

To ease the notation we shall from now on write ξ_n for the n-tuple $(\tau_1, \ldots, \tau_n)$ and thus for instance write $G_{n\xi_n}(t)$ for the right hand side of (3.2).

3.3. Example. For the Poisson process Π_μ (Example 2.4), the waiting times $\sigma_n = \tau_n - \tau_{n-1}$ between jumps are independent and identically exponentially distributed with rate μ. Thus

$$G_{nt_1 \ldots t_n}(t) = e^{-\mu(t-t_n)} \qquad (t > t_n). \qquad \blacksquare$$

3.4. Example. In order that a CCP be stationary Markov it is necessary and sufficient that all waiting times $\sigma_n = \tau_n - \tau_{n-1}$ be independent and exponential. Thus, to each state $n \in \mathbb{N}_0$ there corresponds a rate $\mu_n \geq 0$ such that

$$G_{n t_1 \ldots t_n}(t) = e^{-\mu_n(t-t_n)} \qquad (t > t_n).$$

If at least one $\mu_n = 0$, the process is absorbed at some stage. There-fore, if n_0 is the smallest n such that $\mu_n = 0$, the process is completely specified by the rates $\mu_0, \ldots, \mu_{n_0-1} > 0$, $\mu_{n_0} = 0$. The process will eventually reach n_0 and then remain there forever.

Such a stationary Markov counting process P is either stable or totally unstable, i.e. either $P(\tau_\infty = \infty) = 1$ or $P(\tau_\infty < \infty) = 1$. Fur-thermore, the process is stable iff $\sum \frac{1}{\mu_n} = \infty$. To see this observe that if $\sum \frac{1}{\mu_n} < \infty$, then all $\mu_n > 0$, hence no state is absorbing and $P\tau_\infty = P\Sigma\sigma_n = \Sigma P\sigma_n = \Sigma \frac{1}{\mu_n} < \infty$ so $P(\tau_\infty < \infty) = 1$. If conversely $\Sigma \frac{1}{\mu_n} = \infty$, then either $\mu_n = 0$ for some n in which case the process is absorbed and trivially stable, or else $\mu_n > 0$ for all n and then $Pe^{-\tau_\infty} = P\Pi e^{-\sigma_n} = \Pi Pe^{-\sigma_n} = \Pi(1 + \frac{1}{\mu_n})^{-1} = 0$ wherefore $P(\tau_\infty = \infty) = 1$. ∎

3.5. <u>Example</u>. The generalization from the previous example to Markov CCP's with non-stationary transition probabilities is obtained by attaching to each state $n \in \mathbb{N}_0$ an arbitrary probability on $(0,\infty]$, the survivor function of which we shall denote by $G^{(n)}$. To avoid technical mess we shall assume that the sequence $(t_n^\dagger)$ of termination points for the $G^{(n)}$ is non-decreasing with strict inequality $t_n^\dagger < t_{n+1}^\dagger$ if $t_n^\dagger < \infty$ and $G^{(n)}$ has an atom at $t_n^\dagger$. Then the Markov counting process is given by

$$G_{n t_1 \ldots t_n}(t) = \frac{G^{(n)}(t)}{G^{(n)}(t_n)} \qquad (t > t_n)$$

provided $G^{(n)}(t_n) > 0$. But this is enough to generate the counting process: the conditions on the $t_n^\dagger$ ensure inductively that in the suc-cessive construction of the conditional distributions of τ_n given $\tau_1, \ldots, \tau_{n-1}$ one has $P(\tau_n \leq t_{n-1}^\dagger) = 1$ with $P(\tau_n < t_{n-1}^\dagger) = 1$ if $t_{n-1}^\dagger$ is not an atom for $G^{(n-1)}$, and therefore also that

$P(G^{(n)}(\tau_n) > 0) = 1$.

The Markov property is partly reflected in the observation that $P(\tau_{n+1} > t | F_{\tau_n}) = G^{(n)}(t)/G^{(n)}(\tau_n)$, which shows that for this particular conditional probability, of the past the process remembers only the present. The Markov property is even better understood from (3.10) below, when inserting there the expression for $G_{N_t, \xi(N_t)}$. For a detailed proof of the Markov property, see Jacobsen (1972).

If $G^{(n)}$ has intensity function $\mu^{(n)}$ so that $G^{(n)}(t) = \exp(-\int_0^t ds\mu^{(n)}(s))$, then $t_n^{\dagger}$ is not an atom for $G^{(n)}$ except possibly if $t_n^{\dagger} = \infty$, so we need only assume that $(t_n^{\dagger})$ is non-decreasing. Then

$$G_{nt_1\ldots t_n}(t) = \exp\left(-\int_{t_n}^t ds\mu^{(n)}(s)\right) \qquad (t > t_n). \qquad \blacksquare$$

3.6. **Example**. Consider a probability on $(0,\infty]$ with survivor function G. The renewal sequence corresponding to G is obtained as $(0, S_1, S_1+S_2, \ldots, S_1+\ldots+S_n, \ldots)$ where the S_n are i.i.d. random variables with survivor function G. The renewal sequence determines a counting process K by $K_t = \sup\{n: S_1+\ldots+S_n \leq t\}$, the number of renewals before t. The CCP generated by K is given by

$$G_{nt_1\ldots t_n}(t) = G(t-t_n) \qquad (t > t_n).$$

Examples 3.5 and 3.6 (generalized to describe all Markov CCP) may be combined to show that a CCP is a Markov process and a renewal process iff it is Poisson with constant intensity. $\blacksquare$

3.7. **Example**. Suppose that in Example 3.5 all $G^{(n)} = G$ with G continuous on $(0,\infty)$, so that the assumption on the $t_n^{\dagger} = t^{\dagger}$ is trivially satisfied. The corresponding Markov CCP, P, is then an _inhomogeneous Poisson process_, i.e. it has independent increments and

$$P(N_u - N_t = n | F_t) = \frac{1}{n!}[\log(G(t)/G(u))]^n \frac{G(u)}{G(t)} \quad (n \in \mathbb{N}_0, t \leq u < t^{\dagger}).$$

Also $P(\tau_\infty = t^\dagger) = 1$ if $t^\dagger < \infty$.

For an easy proof, start with the canonical Poisson process Π_1 with rate 1 (Example 2.4). Then define $K_u = N(- \log G(u))$ if $u < t^\dagger$, $K_u = \infty$ if $u \geq t^\dagger$, and argue that the CCP generated by K is P.   ∎

3.8. <u>Example</u>. If in Example 2.6 the X_i are i.i.d. with a survivor function G having no atoms, then for the CCP determined by K

$$G_{nt_1 \ldots t_n}(t) = \left(\frac{G(t)}{G(t_n)}\right)^{r-n} \qquad (t > t_n, \ n \leq r).$$

Since this is a special case of Example 3.5 (with $G^{(n)}(t) = G^{r-n}(t)$), the process is Markov. The expression above for $G_{nt_1 \ldots t_n}$ is most easily found using the original X_i: if $X_{(1)} < \ldots < X_{(r)}$ are the X_i ordered, then

$$G_{nt_1 \ldots t_n}(t) = \mathbb{P}(X_{(n+1)} > t \mid X_{(1)} = t_1, \ldots, X_{(n)} = t_n)$$

and this conditional probability may be found by conditioning on n specific X_i, $X_1, \ldots, X_n$ say, being the smallest and taking the values $t_1, \ldots, t_n$. Writing $X^* = \min(X_{n+1}, \ldots, X_r)$, for $t > t_n$ we then find

$$\mathbb{P}(X^* > t \mid X_1 = t_1, \ldots, X_n = t_n, \ X^* > X_1, \ldots, X_n)$$

$$= \mathbb{P}(X^* > t \mid X_1 = t_1, \ldots, X_n = t_n)/\mathbb{P}(X^* > t_n \mid X_1 = t_1, \ldots, X_n = t_n)$$

$$= \left(\frac{G(t)}{G(t_n)}\right)^{r-n}.$$

Using the same kind of argument, it is an easy exercise to show that the transition probabilities for this Markovian CCP are binomial ones:

$$P_{mn}(t,u) = \binom{r-m}{n-m}\left(1 - \frac{G(u)}{G(t)}\right)^{n-m}\left(\frac{G(u)}{G(t)}\right)^{r-n}$$

for $t \leq u$, $m \leq n \leq r$.   ∎

We shall conclude this section with some results that will prove useful later. For the terminology and results on conditional probabilities used below, see Appendix 1.

3.9. Proposition. Let P be a CCP. For any $t \geq 0$ a regular conditional probability of P given F_t is determined as follows: for any $n \in \mathbb{N}_0$, on the F_t-measurable set $(N_t = n)$,

$$P(\cdot | F_t) = P(\cdot ; \tau_{n+1} > t | F_{\tau_n}) / P(\tau_{n+1} > t | F_{\tau_n}) .$$

Proof. To condition on F_t amounts to conditioning on the number N_t of jumps in $(0,t]$ and the location of these jumps. Since $(N_t = n) = (\tau_n \leq t < \tau_{n+1})$ we therefore find that within $(N_t = n)$,

$$P(\cdot | F_t) = P(\cdot | \xi_n, \tau_n \leq t < \tau_{n+1})$$

$$= P(\cdot ; \tau_{n+1} > t | \xi_n, \tau_n \leq t) / P(\tau_{n+1} > t | \xi_n, \tau_n \leq t) . \qquad \blacksquare$$

The point of the assertion in the proposition is that it reduces the problem of conditioning on F_t to that of conditioning on F_{τ_n} which from the point of view of Theorem 3.1 is more natural.

As an application, let $t \geq 0$ and denote by $\tau_{t,1}$ the time of the first jump after time t: $\tau_{t,1} = \inf\{u > t : N_u \neq N_t\}$. Then on $(N_t = n)$, for $u \geq t$

$$P(\tau_{t,1} > u | F_t) = P(\tau_{n+1} > u | F_{\tau_n}) / P(\tau_{n+1} > t | F_{\tau_n})$$

and consequently

(3.10) $P(\tau_{t,1} > u | F_t) = G_{N_t, \xi_{N_t}}(u) / G_{N_t, \xi_{N_t}}(t) \qquad (u \geq t)$

on $(N_t < \infty)$.

Consider now the conditonal probability $P(\cdot | F_{\tau_n})(w)$ evaluated for a particular path w. This conditional probability freezes the times of the n first jumps at the values $\tau_1(w), \ldots, \tau_n(w)$ and then

generates the remaining jumps according to the recipe of Theorem 3.1.
These jumps themselves determine a counting process on $[\tau_n(w),\infty)$,
which in terms of the original process is simply $N^* = (N_u^*)_{u \geq \tau_n(w)}$,
where $N_u^* = N_u - N_{\tau_n}(w) = N_u - n$, and this new process may therefore
be described by the distribution of the time of its fint jump, the conditional distribution of the time of its second jump given the time of
the first etc. Since the time of the m'th jump of the new process is
the time of the m+n'th jump of the old process, the conditional distribution of the time of the m+1'st jump of the new process given the ti-
mes of the m previous ones is (writing $P^{F_{\tau_m} \text{ at } w}$ for $P(\cdot | F_{\tau_n})(w)$)

$$P^{F_{\tau_m} \text{ at } w}(\tau_{m+1+n} > t | \tau_{n+1}, \ldots, \tau_{n+m})$$

which by the principle of repeated conditioning (see Appendix 1) becomes

$$P(\tau_{m+1+n} > t | \xi_n = (\tau_1(w), \ldots, \tau_n(w)), \tau_{n+1}, \ldots, \tau_{n+m})$$

$$= G_{n+m, \xi_n(w), \tau_{n+1} \cdots \tau_{n+m}}(t).$$

We shall now generalize this to the situation where for an arbitrary
stopping time σ one considers the counting process $N^* = (N_u^*)_{u \geq \sigma}$
beyond σ given by $N_u^* = N_u - N_\sigma$ generated by the conditional proba-
bility $P(\cdot | F_\sigma)$, and then describes N^* by the $P(\cdot | F_\sigma)$ conditional
distribution of the time of the n+1'st jump of N^* given the times of
the n previous ones.

Let σ be a stopping time. For $n \in \mathbb{N}$, let $\tau_{\sigma,n}$ denote the
time of the n'th jump after σ, with the convention that $\tau_{\sigma,n} = \infty$ if
there are less than n jumps after σ. This definition makes sense on
the set $(\sigma < \infty)$. On $(\sigma = \infty)$ simply define all $\tau_{\sigma,n} = \infty$. Note that
on $(\sigma < \infty)$, $\tau_{\sigma,n} = \tau_{N_\sigma + n}$.

3.11. <u>Lemma</u>. For all stopping times σ and all $n \in \mathbb{N}$, $\tau_{\sigma,n}$ is a
stopping time.

Proof. Use the identity

$$(\tau_{\sigma,n} \leq t) = (\sigma \leq t, \; N_\sigma < \infty, \; N_t - N_\sigma \geq n)$$

and observe (for instance by an equivalence class argument) that the set on the right hand side belongs to F_t. ▌

3.12. Theorem. Let P be a CCP with $(G_{nt_1 \ldots t_n})$ the family of survivor functions determining the conditional jump time distributions as in (3.2). Then for any $t \geq 0$ and any $n \in \mathbb{N}_0$, with respect to the conditional probability $P^{F(t)} = P(\cdot | F_t)$, the conditional distribution of $\tau_{t,n+1}$ given $\tau_{t,1}, \ldots, \tau_{t,n}$ is given by

$$P^{F(t)}(\tau_{t,n+1} > u | \tau_{t,1}, \ldots, \tau_{t,n}) = \begin{cases} \dfrac{G_{N_t, \xi_{N_t}}(u)}{G_{N_t, \xi_{N_t}}(t)} & (u \geq t, \quad n = 0) \\[2mm] G_{N_t+n, \xi_{N_t}+n}(u) & (u \geq \tau_{t,n}, n \geq 1). \end{cases}$$

More generally, if σ is a stopping time, then for any $n \in \mathbb{N}_0$, with respect to the conditional probability $P^{F(\sigma)} = P(\cdot | F_\sigma)$ considered on the set $(\sigma < \infty)$, the conditional distribution of $\tau_{\sigma,n+1}$ given $\tau_{\sigma,1}, \ldots, \tau_{\sigma,n}$ is given by

$$(3.13) \quad P^{F(\sigma)}(\tau_{\sigma,n+1} > u | \tau_{\sigma,1}, \ldots, \tau_{\sigma,n}) = \begin{cases} \dfrac{G_{N_\sigma, \xi_{N_\sigma}}(u)}{G_{N_\sigma, \xi_{N_\sigma}}(\sigma)} & (u \geq \sigma, \; n = 0) \\[2mm] G_{N_\sigma+n, \xi_{N_\sigma}+n}(u) & (u \geq \tau_{\sigma,n}, \; n \geq 1). \end{cases}$$

Proof. By the principle of repeated conditioning, finding with respect to $P^{F(\sigma)}$ the conditional distribution of $\tau_{\sigma,n+1}$ given $\tau_{\sigma,1}, \ldots, \tau_{\sigma,n}$ amounts to finding with respect to P the conditional distribution of $\tau_{\sigma,n+1}$ given the σ-algebra generated by F_σ and $\tau_{\sigma,1}, \ldots, \tau_{\sigma,n}$ which is exactly the σ-algebra $F_{\tau_{\sigma,n}}$. Since $\tau_{\sigma,n+1} = \tau_{\tau_{\sigma,n},1}$, the theorem will therefore follow if we show that for any stopping time σ

$$(3.14) \qquad P(\tau_{\sigma,1} > u | F_\sigma) = \frac{G_{N_\sigma,\xi_{N_\sigma}}(u)}{G_{N_\sigma,\xi_{N_\sigma}}(\sigma)}$$

on $(\sigma < \infty)$ for $u \geq \sigma$. The argument for this relies on an approximation of σ by a decreasing sequence of stopping times taking only
countably many values. For $m \in \mathbb{N}$ define

$$\sigma^{(m)} = \sum_{k=1}^{\infty} \frac{k}{2^m} 1_{(\frac{k-1}{2^m} \leq \sigma < \frac{k}{2^n})} + \infty \cdot 1_{(\sigma=\infty)} \ .$$

Then each $\sigma^{(m)}$ is a stopping time $((\sigma^{(m)} = \frac{k}{2^m}) = (\frac{k-1}{2^m} \leq \sigma < \frac{k}{2^m}) \in F_{k/2^m}$
by Proposition 2.9) and $\sigma^{(m)} \downarrow \sigma$. Now let $F \in F(\sigma < \infty)$, $u \geq 0$.
Since

$$\lim_{m \to \infty} 1_{(\tau_{\sigma^{(m)},1} > u)} = 1_{(\tau_{\sigma,1} > u)}$$

it follows by dominated convergence that

$$P(\tau_{\sigma,1} > u; F) = \lim_{m \to \infty} P(\tau_{\sigma^{(m)},1} > u; F)$$

$$= \lim_{m \to \infty} \sum_{k=1}^{\infty} P(\tau_{\frac{k}{2^m},1} > u; F(\frac{k-1}{2^m} \leq \sigma \leq \frac{k}{2^m})) \ .$$

Now $F(\frac{k-1}{2^m} \leq \sigma < \frac{k}{2^m}) \in F_{k/2^m}$, so conditioning on $F_{k/2^m}$ and using
(3.10) reduces the sum to

$$(3.15) \qquad P \sum_{k:\frac{k}{2^m} \leq u} \frac{G(u)}{G(\frac{k}{2^m})} 1_{F(\frac{k-1}{2^m} \leq \sigma < \frac{k}{2^m})} \ ,$$

where $G(t)$ for the k'th term is $G_{N(k2^{-m}),\xi_{N(k2^{-m})}}$.

But using that for any path $w \in (\sigma < \infty)$, if for every m, k
is determined so that $\frac{k-1}{2^m} \leq \sigma(w) < \frac{k}{2^m}$, then $N_{\frac{k}{2^m}}(w) = N_\sigma(w)$ for m
sufficiently large, it follows that the integrand in (3.15) converges
pointwise to

$$[G_{N_\sigma,\xi_{N_\sigma}}(u)/G_{N_\sigma,\xi_{N_\sigma}}(\sigma)] 1_{F(\sigma \leq u)}$$

and (3.14) then follows using dominated convergence. ▮

For $n = 0$, the left hand side of (3.13) is of course $P^{F(\sigma)}(\tau_{\sigma,1} > u)$. Notice that on the right hand side, the denominator is 1 if σ is the time of a jump, so that $\tau_{N_\sigma} = \sigma$.

In the formulation of Theorem 3.12 it was tacitly assumed that there is a regular conditional probability $P^{F(\sigma)}$ of P given F_σ. This follows from standard existence theorems on regular conditional probabilities. However, (3.13) may actually be used for the construction of $P^{F(\sigma)}$: fixing $w \in \overline{W}$ with $\sigma(w) < \infty$, apply (3.13) to generate jump times $\tau_{\sigma,1}, \tau_{\sigma,2}, \ldots$ with

$$\frac{G_{N_\sigma(w),\xi(N_\sigma(w)(w)}(u)}{G_{N_\sigma(w),\xi_{N_\sigma(w)}(w)}(\sigma(w))}$$

the survivor function of $\tau_{\sigma,1}$ and

$$G_{N_\sigma(w)+n,\xi(N_\sigma(w)(w),\tau_{\sigma,1}\cdots\tau_{\sigma,n}}(u)$$

the conditional survivor function of $\tau_{\sigma,n+1}$ gives $\tau_{\sigma,1},\ldots,\tau_{\sigma,n}$. Then the CCP generated by the jump times $\tau_1(w),\ldots,\tau_{N_\sigma(w)}(w),\tau_{\sigma,1}$, $\tau_{\sigma,2},\ldots$ is the regular conditional probability $P^{F(\sigma)}$ at $w = P(\cdot|F_\sigma)(w)$. It obviously has the property that for each w, $P^{F(\sigma)}$ at w is concentrated on the F_σ-atom containing w.

1.4 Intensities for canonical counting processes.

For the Poisson process Π_μ with constant intensity μ the limits

$$\lim_{h\downarrow\downarrow 0} \frac{1}{h} \Pi_\mu(N_{t+h}-N_t \geq 1|F_t) = \lim_{h\downarrow\downarrow 0} \frac{1}{h}(1-e^{-\mu h})$$

$$\lim_{h\downarrow\downarrow 0} \frac{1}{h} \Pi_\mu(N_{t+h}-N_t = 1|F_t) = \lim_{h\downarrow\downarrow 0} \frac{1}{h}\mu he^{-\mu h}$$

$$\lim_{h\downarrow\downarrow 0} \frac{1}{h} \Pi_\mu(N_{t+h}-N_t|F_t) = \lim_{h\downarrow\downarrow 0} \frac{1}{h}\mu h$$

all exist and equal μ. We shall now discuss a class of CCP's for which the limit

$$\lim_{h\downarrow\downarrow 0} \frac{1}{h} P(N_{t+h}-N_t \geq 1|F_t)$$

always exists, and also see what happens to the two other types of limits.

Suppose that the jump time distributions are specified as in (3.2):

$$P(\tau_{n+1} > t|F_{\tau_n}) = G_{n\xi_n}(t) \qquad (t \geq \tau_n).$$

The basic assumption we shall make is that all the survivor functions $G_{nt_1\ldots t_n}$ have a smooth denisty with intensity

$$\mu_{nt_1\ldots t_n}(t) = D^+(-\log G_{nt_1\ldots t_n})(t) \qquad (t \geq t_n)$$

which is right-continuous with left-limits, cf. Section 1.1. Thus

$$G_{nt_1\ldots t_n}(t) = \exp(-\int_{t_n}^t ds\mu_{nt_1\ldots t_n}(s)) \qquad (t \geq t_n),$$

and (3.10) may be written

(4.1) $$P(\tau_{t,1} > u|F_t) = \exp(-\int_t^u ds\mu_{N_t\xi_{N_t}}(s)) \qquad (u \geq t)$$

on $(N_t < \infty)$.

We shall denote by H <u>the class of CCP's for which all</u> $G_{nt_1\ldots t_n}$ <u>have smooth densities</u>, and shall as above write $\mu_{nt_1\ldots t_n}$ for the intensities.

Suppose that P is of class H and consider the stochastic process $(\lambda_t)_{t \geq 0}$ on $(\overline{W}, F)$ given by

$$\lambda_t = \begin{cases} \mu_{N_t \xi_{N_t}}(t) & \text{on } (N_t < \infty) \\ 0 & \text{on } (N_t = \infty). \end{cases}$$

4.2. **Proposition.** The process λ is adapted to (F_t) and has sample paths which are everywhere right-continuous and, P-almost surely, have left-limits on $(0, \tau_\infty)$. Further, the paths are right locally integrable in the sense that for all $t \geq 0$, P-almost all $w \in \overline{W}$, $\int_t^{t+h} ds \, \lambda_s(w) < \infty$ for $h > 0$ sufficiently small. Finally, for every $t \geq 0$

$$\lim_{h \downarrow \downarrow 0} \frac{1}{h} P(N_{t+h} - N_t \geq 1 | F_t) = \lambda_t$$

P-a.s. on $(N_t < \infty)$.

Proof. It is clear that for every t, $w \to \lambda_t(w)$ is constant on each F_t-atom. But this mapping is F-measurable because on $(N_t = n)$ it agrees with the mapping $w \to \mu_{n\xi_n(w)}(t)$ which is composed from the two measurable mappings $w \to \xi_n(w)$ and $(t_1, \ldots, t_n) \to \mu_{nt_1 \ldots t_n}(t)$. Thus λ is (F_t)-adapted.

As intensities the functions $t \to \mu_{nt_1 \ldots t_n}(t)$ have all the properties described in Section 1.1 for an intensity μ. It should therefore be clear that for a given w, $t \to \lambda_t(w)$ has left-limits on $(0, \infty)$ and is right locally integrable provided $\tau_{n+1}(w) < \tau_n^\dagger(w)$ for every $n \geq 0$, writing $\tau_n^\dagger(w)$ for the termination point of the distribution with intensity $\mu_{n\xi_n(w)}$. But (3.2) and the fact that $G_{n\xi_n}$ has no atoms implies that $P(\tau_{n+1} < \tau_n^\dagger) = 1$, so we have shown that P-almost surely, $t \to \lambda_t$ has left-limits on $(0, \infty)$ and is right locally integrable. The right-continuity of λ is obvious.

As for the proof of the last assertion, observe that by (4.1)

$$P(N_{t+h}-N_t \geq 1 | F_t) = P(\tau_{t,1} \leq t+h | F_t)$$

$$= 1-\exp(-\int_t^{t+h} ds \; \mu_{N_t \xi_{N_t}}(s))$$

on $(N_t < \infty)$. Now simply use the right-continuity of $\mu_{N_t \xi_{N_t}}$ to obtain the desired limit.

If P of class H is given, the $\mu_{n t_1 \ldots t_n}$ are of course not uniquely determined from P, since they are intensities for conditional probabilities with respect to t. However, it is certainly true that with $P^{(n)}$ the P-distribution of $(\tau_1, \ldots, \tau_n)$ the right-continuous functions $t \to \mu_{n t_1 \ldots t_n}(t)$ are, for every fixed n, uniquely determined except for a set of $(t_1, \ldots, t_n)$ of $P^{(n)}$-probability 0. Translating this into a statement about the right-continuous process λ, we see that λ is not determined uniquely from P, but that any two versions λ and λ' will have the property that for P-almost all w, $\lambda_t(w) = \lambda_t'(w)$ simultaneously for all $t \geq 0$: in the terminology of the general theory of processes, λ and λ' are <u>indistinguishable</u>. Henceforth, when deriving λ from P, we mean of course that λ is one of the possible versions.

These considerations lead to the following very important definition, which is stated modulo indistinguishability.

4.3. <u>Definition</u>. For a canonical counting process P of class H, the <u>intensity process</u> $\lambda_- = (\lambda_{t-})_{t>0}$ is given by $\lambda_{t-} = \lim_{s \uparrow\uparrow t} \lambda_s$ for $t < \tau_\infty$, $\lambda_{t-} = 0$ for $t \geq \tau_\infty$. ▎

The properties of λ discussed above yield the following properties of λ_-: the intensity process is defined almost surely, is left-continuous on $(0, \tau_\infty)$ with right-limits everywhere on $[0, \infty)$, right locally integrable and saitsfies $\lambda_{t-} = 0$ for $t \geq \tau_\infty$.

One reason for using λ_- as intensity process rather than λ

appear in Proposition 4.8 (e) below. From the point of view of general process theory the choice of λ_- has to do with the desire to use only predictable intensities. Predictable processes are treated in Chapter 3.

Comparing the last assertion of Proposition 4.2 with the expressions for the Poisson intensity given at the beginning of this section, it should be observed that it is not in general true that

$$\lim_{h\downarrow\downarrow 0} \frac{1}{h} P(N_{t+h} - N_t = 1 | F_t) = \lambda_t .$$

For this to hold, one must have that

$$\lim_{h\downarrow\downarrow 0} \frac{1}{h} P(N_{t+h} - N_t \geq 2 | F_t) = 0$$

and counterexamples to this are easily constructed: consider $t = 0$ and take $\mu_0 = 1$, $\mu_{1,s}(u) = \varphi(s)$ with φ increasing to ∞ sufficiently rapidly as $s \downarrow\downarrow 0$. For the same reason it is not true that

$$\lim_{h\downarrow\downarrow 0} \frac{1}{h} P(N_{t+h} - N_t | F_t) = \lambda_t ,$$

not even if all N_t have finite expectations. Notice that always

$$P(N_{t+h} - N_t = 1 | F_t) \leq P(N_{t+h} - N_t \geq 1 | F_t) \leq P(N_{t+h} - N_t | F_t) .$$

The properties of λ_- listed below Definition 4.3 characterizes intensity processes as we shall now show.

4.4. Proposition. Suppose $\lambda_- = (\lambda_{t-})_{t>0}$ is a non-negative F_t-adapted process with sample paths which are left-continuous on $(0, \tau_\infty)$ with right-limits everywhere, locally right integrable and such that $\lambda_{t-} = 0$ for $t \geq \tau_\infty$. Then there is a unique CCP of class H which has λ_- as intensity process.

Proof. Denote by λ the right-continuous regularisation of λ_-, i.e. $\lambda_t = \lim_{s\downarrow\downarrow t} \lambda_{s-}$ for $t \geq 0$. Then P of class H certainly has intensity λ_- if for all $w \in \overline{W}$, $t \geq 0$ with $N_t(w) < \infty$

$$\mu_{N_t(w)}, \xi_{N_t}(w) = \lambda_t(w) .$$

But from this all the intensities $\mu_{nt_1 \cdots t_n}$ of the conditional jump time distributions can be determined completely, viz. for $n \in \mathbb{N}_0$, $t_1 < \cdots < t_n \leq t$ define

$$\mu_{nt_1 \cdots t_n}(t) = \lambda_t(w)$$

for any w such that $N_t(w) = n$, $\tau_1(w) = t_1, \cdots, \tau_n(w) = t_n$. The definition in unambiguous because λ_t is F_t-measurable, the result is a measurable function of $(t_1, \cdots, t_n)$ and $\mu_{nt_1 \cdots t_n}$ is right-continuous with left-limits because $t \to \lambda_t(w)$ is. Finally (and this is important) $\mu_{nt_1 \cdots t_n}$ is the intensity of a probability on $(t_n, \infty]$ because $\int_{t_n}^{t_n+h} ds\, \mu_{nt_1 \cdots t_n}(s) < \infty$ for $h > 0$ sufficiently small by the right local integrability of λ.

That this P is unique may be argued as follows: suppose P' is a second CCP of class H with intensity λ_- and suppose that $P \neq P'$. Then there is a smallest n such that the P-distribution $P^{(n)}$ of $(\tau_1, \cdots, \tau_n)$ is different from the P'-distribution $P'^{(n)}$ of $(\tau_1, \cdots, \tau_n)$ and consequently $P^{(n-1)} = P'^{(n-1)}$ and the set A of $(t_1, \cdots, t_{n-1})$ such that $\mu_{n-1, t_1, \cdots, t_{n-1}} \neq \mu'_{n-1, t_1, \cdots, t_{n-1}}$ (as functions) has strictly positive $P^{(n-1)}$ and $P'^{(n-1)}$-measure. But considering now the set $F = ((\tau_1, \cdots, \tau_{n-1}) \in A)$ it is clear that $P(F) = P'(F) > 0$ while the intensities for P and P' in at least one pair of versions are different everywhere on F. Since a CCP determines its intensity up to indistinguishability, we have reached a contradiction. ▌

In terms of the intensity process it is possible to give a simple sufficient condition for a CCP of class H to be stable. The condition below can be improved by imposing some obvious "almost surely" statements.

4.5. Proposition. In order that a CCP of class H be stable it is sufficient that there exist a sequence $(\kappa_n)_{n \geq 0}$ of non-negative constants with $\Sigma \frac{1}{\kappa_n} = \infty$ such that

$$\lambda_t \leq \kappa_{N_t} \qquad\qquad (t \geq 0)$$

or, equivalently that

$$\mu_{nt_1 \cdots t_n}(t) \leq \kappa_n \qquad\qquad (n \in \mathbb{N}_0, \, t_1 < \cdots < t_n \leq t).$$

Proof. By the definition of the intensity process, the two conditions are obviously identical, so we need only show that the second implies stability.

Let P be the (stationary) Markov CCP for which the waiting times $\sigma_n = \tau_n - \tau_{n-1}$ are independent, σ_n being exponential with rate κ_{n-1}. Then according to Example 3.4, P is stable.

For $n \in \mathbb{N}_0$, $t_1 < \cdots < t_n$, $t \geq 0$ define

$$H_{nt_1 \cdots t_n}(t) = \frac{1}{\kappa_n} \int_0^t ds \, \mu_{nt_1 \cdots t_n}(t_n + s)$$

with $H_{nt_1 \cdots t_n} \equiv \infty$ if $\kappa_n = 0$, and let $H^{-1}_{nt_1 \cdots t_n}$ denote the right-continuous inverse of $H_{nt_1 \cdots t_n}$:

$$H^{-1}_{nt_1 \cdots t_n}(u) = \inf\{t > 0 \colon H_{nt_1 \cdots t_n}(t) > u\}.$$

The function $H^{-1}_{nt_1 \cdots t_n}$ is defined on $[0, H_{nt_1 \cdots t_n}(\infty))$ where $H_{nt_1 \cdots t_n}(\infty) = \frac{1}{\kappa_n} \int_0^\infty ds \, \mu_{nt_1 \cdots t_n}(s)$. We define it on all of $[0,\infty)$ by putting it equal to ∞ on $[H(\infty),\infty)$.

Now consider the sequence $(T_n)_{n \geq 0}$ of random variables given by $T_0 = 0$ and

$$T_{n+1} = T_n + H^{-1}_{nT_1 \cdots T_n}(\sigma_{n+1}).$$

Since all σ_n may be thought of as being strictly positive, it is clear from the definition of the inverse functions that $T_{n+1} > T_n$ whenever $T_n < \infty$. Thus the T_n may be taken as the sequence of jump times for a new counting process for which the conditional jump time distributions are, on $(T_n < \infty)$ for $t \geq T_n$

$$P(T_{n+1} \geq t | T_1, \cdots, T_n) = P(H_{nT_1 \cdots T_n}^{-1}(\sigma_{n+1}) \geq t - T_n | T_1, \cdots, T_n)$$

$$= P(\sigma_{n+1} \geq H_{nT_1 \cdots T_n}(\infty) | T_1, \cdots, T_n)$$

$$+ P(H_{nT_1 \cdots T_n}^{-1}(\sigma_{n+1}) \geq t - T_n, \sigma_{n+1} < H_{nT_1, \cdots, T_n}(\infty) | T_1, \cdots, T_n)$$

$$= P(\sigma_{n+1} \geq H_{nT_1 \cdots T_n}(t - T_n) | T_1, \cdots, T_n)$$

using carefully the properties of H^{-1}. The definition of the T_n shows that $(T_1, \cdots, T_n)$ is determined from $(\sigma_1, \cdots, \sigma_n)$ and this vector is therefore independent of σ_{n+1}. Consequently

$$P(T_{n+1} \geq t | T_1, \cdots, T_n) = \exp(-\kappa_n H_{nT_1 \cdots T_n}(t - T_n))$$

$$= \exp\left(- \int_{T_n}^{t} ds \; \mu_{nT_1 \cdots T_n}(s)\right),$$

that is to say that the CCP determined by (T_n) has conditional jump intensities $\mu_{nt_1 \cdots t_n}$.

Finally, since by assumption $H_{nt_1 \cdots t_n}(t) \leq t$ it follows that $T_{n+1} \geq T_n + \sigma_{n+1}$, therefore also $T_n \geq \tau_n$ for all n wherefore P stable forces the counting process with jump times (T_n) to be stable.

❙

Remark. The proof which is a stochastic comparison argument is really based on a converse to Proposition 1.4, or alternatively, to an inverse of the type of time substitution employed in Proposition 4.6 below.

❙

For a CCP of class H with intensity process λ_-, the integrated intensity process is defined as

$$\Lambda_t = \int_0^t ds \; \lambda_{s-} = \int_0^t ds \; \lambda_s \qquad (t \geq 0).$$

Here for every $w \in \overline{W}$ the integrals are ordinary Lebesgue integrals. They are equal because the left-continuity, right-limit properties of λ_- ensure that the sample paths of this process have at most countably many discontinuities.

The integrated intensity has the following properties: the sample paths are continuous and non decreasing, taking values in $[0,\infty]$ with $\Lambda_0 = 0$. The process is F_t-adapted. We shall write $\Lambda_\infty = \lim_{t \uparrow \infty} \Lambda_t$.

For $w \in \overline{W}$, denote by $u \to \rho_u(w)$ the right-continuous inverse of the non-decreasing function $t \to \Lambda_t(w)$:

$$\rho_u(w) = \inf\{t > 0 : \Lambda_t(w) > u\}.$$

Then $(\rho_u \geq t) = (\Lambda_t \leq u)$ so each ρ_u is a stopping time by Proposition 2.9.

4.6. **Proposition.** Suppose that P is a CCP of class H with integrated intensity $(\Lambda_t)_{t \geq 0}$ and let $\mu > 0$ be a constant. Then if $\int_{t_n}^\infty ds \; \mu_{nt_1 \cdots t_n}(s) = \infty$ for all $n \in \mathbb{N}_0$, $0 < t_1 < \cdots < t_n$, the stochastic process N^* defined by

$$N_u^* = N_{\rho_{\mu u}} \qquad (u \geq 0)$$

is a Poisson process with intensity μ.

Proof. The assumption on the intensities guarantees that the process P is not absorbed so that $P(\tau_n < \infty) = 1$ for all n. Furthermore, from Proposition 4.8 below it follows that $\Lambda(\tau_{n+1}) > \Lambda(\tau_n)$ a.s. with $\Lambda(\tau_n) < \infty$ and increasing to ∞ a.s. as $n \uparrow \infty$. Therefore

$$(N_u^* = n) = (\tau_n \leq \rho_{\mu u} < \tau_{n+1}) = (\Lambda(\tau_n) \leq \mu u < \Lambda(\tau_{n+1}))$$

so the jump times for N^* is the sequence (τ_n^*) where $\tau_n^* = \frac{1}{\mu}\Lambda(\tau_n)$, and N^* is indeed a counting process such that $P(N_u^* < \infty) = 1$ for all $u \geq 0$ and $N_\infty^* = \infty$ a.s. The proposition will therefore follow if we show that under P the waiting times $\tau_1^*, \tau_2^* - \tau_1^*, \cdots$ are i.i.d exponential at rate μ. Since τ_n^* is F_{τ_n}-measurable, for this it is enough to show that

(4.7) $$P(\tau_{n+1}^* - \tau_n^* > v | F_{\tau_n}) = e^{-\mu v} \quad (n \in \mathbb{N}_0, v \geq 0).$$

Now, on the set $(\tau_n^* < \infty)$ (which has full measure)

$$\tau_{n+1}^* - \tau_n^* = \frac{1}{\mu}\int_{\tau_n}^{\tau_{n+1}} ds\ \lambda_s = \frac{1}{\mu}\int_0^{\tau_{n+1}-\tau_n} ds\ \mu_{n\tau_1\cdots\tau_n}(s+\tau_n),$$

and since $s \to \mu_{n\tau_1\cdots\tau_n}(s+\tau_n)$ is the intensity of the conditional distribution of $\tau_{n+1} - \tau_n$ given F_{τ_n}, (4.7) follows from Proposition 1.4. ❙

The final result of this section summarizes some important properties of the integrated intensity. Part b) of Propositon 4.8 is theoretically extremely relevant, but not of much practical use as a criterion for stability. (It cannot be used to give the stability criterion of Example 3.4 for instance).

4.8. Proposition. Let P be a CCP of class H. Then

(a) For all $n \in \mathbb{N}_0$, $\Lambda(\tau_n) < \Lambda(\tau_{n+1})$ P-a.s. on $(\Lambda(\tau_n) < \infty, \tau_{n+1} < \infty)$.

(b) One has $\tau^* = \tau_\infty$ P-a.s. where $\tau^* = \inf\{t \geq 0 : \Lambda_t = \infty\}$.

(c) For all $n \in \mathbb{N}_0$, $\Lambda(\tau_n) < \infty$ P-a.s.

(d) One has $\Lambda_\infty = \infty$ P-a.s. on $(N_\infty = \infty)$.

(e) For all $n \in \mathbb{N}_0$, $\lambda_{\tau_n^-} > 0$ P-a.s. on $(\tau_n < \infty)$.

<u>Proof</u>. Throughout the proof we shall write $\mu(u)$ as short for

$\mu_{n\xi_n}(u)$.

For any $n \in \mathbb{N}_0$, on the set $(\Lambda_{\tau_n} < \infty, \ \tau_n < \infty)$

$$P(\Lambda_{\tau_{n+1}} = \Lambda_{\tau_n}, \tau_{n+1} < \infty | F_{\tau_n}) = P\left(\int_{\tau_n}^{\tau_{n+1}} ds \ \mu(s) = 0, \ \tau_{n+1} < \infty | F_{\tau_n}\right)$$

$$= \int_{\tau_n}^{\infty} du \ \mu(u) \exp\left(-\int_{\tau_n}^{u} \mu\right) 1_{\left(\int_{\tau_n}^{u} \mu = 0\right)}$$

which reduces to 0 and proves (a).

Next, observe that on $(\tau_n < \infty)$

(4.9)
$$P(\tau_{n+1} = \infty | F_{\tau_n}) = \exp\left(-\int_{\tau_n}^{\infty} \mu\right)$$

so that on the set $(\tau_n < \tau^*) \in F_{\tau_n}$

$$P(\tau_{n+1} \geq \tau^*, \tau^* < \infty | F_{\tau_n}) \leq P\left(\int_{\tau_n}^{\tau_{n+1}} \mu = \infty | F_{\tau_n}\right)$$

$$= \int_{\tau_n}^{\infty} du \ \mu(u) \exp(-\int_{\tau_n}^{u} \mu) 1_{\left(\int_{\tau_n}^{u} \mu = \infty\right)} + \exp(-\int_{\tau_n}^{\infty} \mu) 1_{\left(\int_{\tau_n}^{\infty} \mu = \infty\right)}$$

working on $(\tau_{n+1} < \infty)$ and $(\tau_{n+1} = \infty)$ separately. Again we get 0
so that $\tau_{n+1} < \tau^*$ a.s. on $(\tau_n < \tau^* < \infty)$. For $n = 0$ this gives
$\tau_1 < \tau^*$ a.s. on $(\tau^* < \infty)$ and by induction therefore $\tau_n < \tau^*$ a.s.
on $(\tau^* < \infty)$ for all n, which immediately implies that $\tau_\infty \leq \tau^*$ a.s.

The opposite inequality is trivial on $(\tau_\infty = \infty)$, and it is there-
fore sufficient to verify that it holds on $A = \bigcap_{n \geq 1} (\tau_n < \infty) \supset (\tau_\infty < \infty)$.
But on $(\tau_n < \infty)$,

$$P(\exp(-\Lambda_{\tau_{n+1}}) ; \tau_{n+1} < \infty | F_{\tau_n})$$

$$= \exp(-\Lambda_{\tau_n}) P(\exp(-\int_{\tau_n}^{\tau_{n+1}} ds \mu(s)) ; \tau_{n+1} < \infty | F_{\tau_n})$$

$$= \exp(-\Lambda_{\tau_n}) \int_{\tau_n}^{\infty} du\,\mu(u)\exp\left(-\int_{\tau_n}^{u}\mu\right)\exp\left(-\int_{\tau_n}^{u}\mu\right)$$

$$= \exp(-\Lambda_{\tau_n})\left[-\tfrac{1}{2}\exp\left(-2\int_{\tau_n}^{u}\mu\right)\right]_{u=\tau_n}^{\infty}$$

$$\leq \tfrac{1}{2}\exp(-\Lambda_{\tau_n}).$$

Therefore

$$P(\exp(-\Lambda_{\tau_{n+1}});\ \tau_{n+1} < \infty) \leq \tfrac{1}{2}\,P(\exp(-\Lambda_{\tau_n});\ \tau_n < \infty)$$

and consequently

$$P(\exp(-\Lambda_{\tau_n});\ \tau_n < \infty) \leq \frac{1}{2^n}$$

for $n \in \mathbb{N}_0$. Letting $n \uparrow \infty$ and using monotone convergence it follows that

$$P(\exp(-\Lambda_{\tau_\infty});\ A) = 0,$$

i.e. $\Lambda_{\tau_\infty} = \infty$ a.s. on A, or equivalently $\tau_\infty \geq \tau^*$ a.s. on A and (b) is proved. Since $A = (N_\infty = \infty)$, (d) follows also.

To prove (c), first note that since $\Lambda_{\tau_n} < \infty$ on $(\tau_n < \tau^*)$, then also $\Lambda_{\tau_n} < \infty$ a.s. on $(\tau_n < \infty)$ since because of (b), $(\tau_n < \infty) = \{\tau_n < \tau_{n+1} \leq \tau^*\} = (\tau_n < \tau^*)$ a.s. To see next that $\Lambda_{\tau_n} < \infty$ a.s. on $(\tau_n = \infty)$, use that on $(\Lambda_{\tau_n} < \infty,\ \tau_n < \infty)$ and due to (4.9)

$$P(\Lambda_{\tau_{n+1}} = \infty,\ \tau_{n+1} = \infty | F_{\tau_n}) = P\left(\int_{\tau_n}^{\tau_{n+1}}\mu = \infty,\ \tau_{n+1} = \infty | F_{\tau_n}\right)$$

$$= 1_{\left(\int_{\tau_n}^{\infty}\mu = \infty\right)}\exp\left(-\int_{\tau_n}^{\infty}\right) = 0$$

so that $\Lambda_{\tau_{n+1}} < \infty$ a.s. on $(\Lambda_{\tau_n} < \infty,\ \tau_n < \tau_{n+1} = \infty)$, which forces $\Lambda_{\tau_n} < \infty$ a.s. on $(\tau_m = \infty)$ for all m.

To establish (e), we find that on $(\tau_n < \infty)$

$$P(\lambda_{\tau_{n+1}-} = 0, \tau_{n+1} < \infty | F_{\tau_n}) = P(\mu(\tau_{n+1}-) = 0,\ \tau_{n+1} < \infty | F_{\tau_n})$$

$$= \int_{\tau_n}^{\infty} ds\,\mu(s)\exp\left(-\int_{\tau_n}^{s}\mu\right)1_{(\mu(s-)=0)}(s) = 0$$

because $\mu(s-) = \mu(s)$ except for countably many values of s. ∎

Notice that it is essential that one considers the left-limit in statement (e): if e.g. $\mu_0 \equiv 1$, $\mu_{1,s} \equiv 0$ for all s, then $P(\lambda_{\tau_1} = 0) = 1$.

We have so far considered intensities for canonical counting processes. If instead we are given a counting process $K = (K_t)_{t \geq 0}$ on a filtered space $(\Omega, A, A_t, \mathbb{P})$ one may discuss (right-continuous) intensities of the form

$$\lambda_t' = \lim_{h \downarrow \downarrow 0} \frac{1}{h} P(K_{t+h} - K_t \geq 1 | A_t).$$

Assuming the limit to exist, λ' becomes an A_t-adapted process, which it is then natural to relate to the (right-continuous) intensity λ of the CCP P generated by K.

Let $T: \Omega \to \bar{W}$ denote the transformation $\omega \to (K_t(\omega))_{t \geq 0}$, so that in particular $P = T(\mathbb{P})$. Then

$$\lambda_t \circ T = \lim_{h \downarrow \downarrow 0} \frac{1}{h} P(N_{t+h} - N_t \geq 1 | \dot{F}_t) \circ T$$
$$= \lim_{h \downarrow \downarrow 0} \frac{1}{h} \mathbb{P}(K_{t+h} - K_t \geq 1 | T^{-1} F_t)$$

and computing this conditional probability by conditioning first on A_t and then on the smaller σ-algebra $T^{-1} F_t$, it is reasonable to expect that

(4.10) $$\lambda_t \circ T = \mathbb{P}(\lambda_t' | T^{-1} F_t).$$

To prove this one must justify why the operations of taking limits as $h \downarrow \downarrow 0$ and performing conditional expectations given $T^{-1} F_t$, can be interchanged. This may be done if e.g. all λ_t' are assumed $\mathbb{P}$-integrable. We shall not go into the details, but only say that (4.10) holds under very reasonable regularity assumptions. The result is known as the _innovation theorem_. It will not be used in these notes.

1.5 Martingale decompositions for canonical counting processes.

Consider the canonical Poisson process Π_μ with constant intensity $\mu > 0$. Then, defining $M = (M_t)_{t \geq 0}$ by $M_t = N_t - \mu t$ it is true that with respect to Π_μ

i) $\qquad\qquad\qquad (M_t, F_t)_{t \geq 0}$ is a martingale

ii) $\qquad\qquad\qquad (M_t^2 - \mu t, F_t)_{t \geq 0}$ is a martingale

as is seen easily using the independent increment property of Π_μ.

We shall now generalize this result to CCP's of class H. Proposition 4.6 suggests that conditionally on F_t a CCP P with integrated intensity Λ, immediately after t behaves like a Poisson process with (constant) intensity λ_t. It is therefore not surprising that for the general result μt should be replaced by Λ_t.

Given a CCP, P, we shall call a process $X = (X_t)_{t \geq 0}$ defined on the path-space W or $\overline{W}$ a P-martingale (P-submartingale) provided X is (F_t)-adapted with right-continuous paths having left-limits everywhere and further $P|X_t| < \infty$ for all t and

$$P(X_t | F_s) = X_s, \quad (P(X_t | F_s) \geq X_s)$$

for $s \leq t$. Recall that under suitable conditions, by the optional sampling theorem this equality (inequality) remains valid when the pair (s,t) is replaced by a pair (σ, τ) of stopping times $\sigma \leq \tau$. For instance it suffices that $\tau \leq t$ for some t, which is all we use below.

We shall say that a CCP, P, has finite expectations locally if $PN_t < \infty$ for all $t \geq 0$. In particular, P is then stable.

5.1. Theorem. Suppose P is a CCP of class H with finite expectations locally and define $M = (M_t)_{t \geq 0}$ by $M_t = N_t - \Lambda_t$, where Λ is the integrated intensity of P. Then with respect to P

(a) $(M_t, F_t)_{t \geq 0}$ is a martingale;

(b) $(M_t^2 - \Lambda_t, F_t)_{t \geq 0}$ is a martingale.

<u>Proof</u>. We shall first argue, that in order to show the assertions of the theorem for a <u>given</u> P, it is enough to show that

(5.2) $PN_{t \wedge \tau_n} = P\Lambda_{t \wedge \tau_n}$ $(t \geq 0, \; n \in \mathbb{N})$

(5.3) $PM_{t \wedge \tau_n}^2 = P\Lambda_{t \wedge \tau_n}$ $(t \geq 0, \; n \in \mathbb{N})$

for <u>all</u> P of class H with finite expectations locally, (still writing Λ for the integrated intensity for P, which of course depends on P). To see this, fix P and observe first that (5.2), (5.3) imply since $PN_t < \infty$ by assumption and $N_{t \wedge \tau_n} \leq N_t$, that $\Lambda_{t \wedge \tau_n}$, $M_{t \wedge \tau_n}$, $M_{t \wedge \tau_n}^2$ are all P-integrable. Letting $n \uparrow \infty$ in (5.2) and using monotone convergence now gives

(5.2)' $PN_t = P\Lambda_t$ $(t \geq 0)$,

in particular Λ_t and M_t are P-integrable. Because P is stable, $\tau_n \uparrow \infty$ P-a.s., hence $M_{t \wedge \tau_n} \to M_t$ P-a.s., so taking limits as $n \to \infty$ in (5.3) and using Fatou's lemma gives

$$PM_t^2 \leq \liminf PM_{t \wedge \tau_n}^2 = \liminf P\Lambda_{t \wedge \tau_n} = P\Lambda_t$$

so M_t^2 is P-integrable. But then also, if it has been shown that M is a P-martingale, M^2 is a P-submartingale and the optional sampling theorem applied to the bounded stopping times $t \wedge \tau_n \leq t$, gives $PM_t^2 \geq PM_{t \wedge \tau_n}^2$ which using (5.3) and comparing with the inequality above yields

(5.3)' $PM_t^2 = P\Lambda_t$ $(t \geq 0)$.

Thus (5.2), (5.3) implies (5.2)', (5.3)' for all P. To deduce from this that M is a P-martingale, observe that by Theorem 3.12 it follows that conditionally on F_s, the counting process $N' = (N'_t)_{t \geq s}$ beoynd s, where $N'_t = N_t - N_s$, is of class H with integrated intensity $(\Lambda'_t)_{t \geq s}$ where $\Lambda'_t = \Lambda_t - \Lambda_s$. (It may be helpful to think of $P(\cdot|F_s)(w)$ for a given w, as a probability on the set of paths w' with $w' \underset{s}{\sim} w$, rather than as a probability on all of W). Applying $(5.2)'$ to the conditional process therefore yields $P(N'_t|F_s) = P(\Lambda'_t|F_s)$, or equivalently

$$P(N_t|F_s) - N_s = P(\Lambda_t|F_s) - \Lambda_s \qquad (s \leq t),$$

which by rearrangement shows that M is a P-martingale and proves assertion (a).

But then also $(5.3)'$ holds for all P, and applying this to the conditional process yields, writing $M'_t = N'_t - \Lambda'_t = M_t - M_s$,

$$(5.4) \qquad P(M'^2_t|F_s) = P(\Lambda'_t|F_s).$$

But because M is a Martingale also

$$P(M'^2_t|F_s) = P(M^2_t - 2M_s M_t + M^2_s|F_s)$$
$$= P(M^2_t|F_s) - M^2_s$$

and together with (5.4) this proves $M^2 - \Lambda$ to be a P-martingale, establishing assertion (b).

It remains to show (5.2) and (5.3). But for this it suffices to show that

$$(5.5) \qquad PN_{t \wedge \tau_1} = P\Lambda_{t \wedge \tau_1} = PM^2_{t \wedge \tau_1} \qquad (t \geq 0).$$

Namely, assuming this to hold for all P of class H with locally finite expectations, fix P and apply (5.5) to the counting process beyond τ_n generated by the conditional probability $P(\cdot|F_{\tau_n})$. For that process, τ_{n+1} is the time of the first jump and hence (5.5) implies (arguing as above, using Theorem 3.12 to give the integrated intensity of the conditional process and working on the sets $(\tau_n \leq t)$

(the interesting part) and $(\tau_n > t)$ (trivial) separately that

$$P(N_{t\wedge\tau_{n+1}} | F_{\tau_n}) - N_{t\wedge\tau_n} = P(\Lambda_{t\wedge\tau_{n+1}} | F_{\tau_n}) - \Lambda_{t\wedge\tau_n} = P(M^2_{t\wedge\tau_{n+1}} | F_{\tau_n}) - M^2_{t\wedge\tau_n}$$

for all $n \geq 1$, $t \geq 0$. Here everything in sight is P-integrable, so taking expectations we get

$$PN_{t\wedge\tau_{n+1}} - PN_{t\wedge\tau_n} = P\Lambda_{t\wedge\tau_{n+1}} - P\Lambda_{t\wedge\tau_n} = PM^2_{t\wedge\tau_{n+1}} - PM^2_{t\wedge\tau_n}$$

and a trivial induction produces (5.2), (5.3) from this and (5.5).

As the last step in the proof of the theorem, (5.5) is established by explicit calculation:

$$P(N_{t\wedge\tau_1}) = P(N_t ; \tau_1 > t) + P(N_{\tau_1} ; \tau_1 \leq t)$$

$$= P(\tau_1 \leq t) = 1 - e^{-\int_0^t \mu_0} \ ;$$

$$P\Lambda_{t\wedge\tau_1} = P(\Lambda_t ; \tau_1 > t) + P(\Lambda_{\tau_1} ; \tau_1 \leq t)$$

$$= \left(\int_0^t ds \ \mu_0(s) \right) e^{-\int_0^t \mu_0} + \int_0^t ds \ \mu_0(s) s^{-\int_0^s \mu_0} \int_0^s du \ \mu_0(u) \ ;$$

$$PM^2_{t\wedge\tau_1} = P((N_t - \Lambda_t)^2 ; \tau_1 > t) + P((N_{\tau_1} - \Lambda_{\tau_1})^2 ; \tau_1 \leq t)$$

$$= \left(\int_0^t ds \ \mu_0(s) \right)^2 e^{-\int_0^t \mu_0} + \int_0^t ds \ \mu_0(s) e^{-\int_0^s \mu_0} \left(1 - \int_0^s du \ \mu_0(u) \right)^2 \ .$$

That the three expressions are identical is most easily verified by observing that they are all 0 for $t = 0$ and have the same derivatives from the right. ▌

Remark. Suppose that $X = (X_t)_{t\geq 0}$ is a submartingale on a filtered space $(\Omega, A, A_t, \mathbb{P})$, i.e. each X_t is $\mathbb{P}$-integrable, A_t-adapted and $\mathbb{P}(X_t | A_s) \geq X_s$ for $s \leq t$. According to the Doob-Meyer decomposition theorem there is a unique predictable increasing process $A = (A_t)_{t\geq 0}$ with $A_0 = 0$ such that $X-A$ is a martingale. Here increasing means adapted with non-decreasing right-continuous sample paths. If the paths are continuous the process is automatically predictable. (We shall discuss predictable processes on counting process path-spaces in Chapter 3).

Now if P is a CCP of class H with locally finite expectations, then $(N_t, F_t)_{t \geq 0}$ is a submartingale with respect to P, (simply because $t \to N_t$ is increasing). Thus, Theorem 5.1 shows that Λ is the predictable increasing process for the submartingales N and $(N-\Lambda)^2$.

For $X = (X_n)_{n \in \mathbb{N}_0}$ a submartingale in discrete time on a filtered space $(\Omega, A, A_n, \mathbb{P})$, the Doob-Meyer decomposition states that there is a unique increasing process $A = (A_n)_{n \in \mathbb{N}_0}$ with $A_0 = 0$ and A_n A_{n-1}-measurable for $n \geq 1$ such that $X-A$ is a martingale. It is easy to prove this and to find that

$$A_n = \sum_{k=0}^{n-1} \mathbb{P}(X_{k+1} - X_k | F_k).$$

In continuous time the analogue of this is the following: in a suitable sense the limits $\lim_{h \downarrow 0} \frac{1}{h} \mathbb{P}(X_{t+h} - X_t | A_t) = a_t$ exist and $A_t = \int_0^t ds \, a_s$. Of course something like this happens in Theorem 5.1(a), cf. Proposition 4.2. ▌

Remark. It is tempting to try and prove Theorem 5.1 from the fact that it is true for Poisson processes, Proposition 4.6 and the optional sampling theorem. The proposition and properties of Poisson processes show that with $G_u = F_{\rho_{\mu u}}$, the process (M_u^*, G_u) is a martingale, where $M_u^* = N_u^* - \mu u$.

Now $(\frac{1}{\mu} \Lambda_t \leq u) = (\rho_{\mu u} \geq t) \in G_u$, so $\frac{1}{\mu} \Lambda_t$ is a stopping time with respect to the filtration (G_u). Therefore, if the optional sampling theorem applies one should have for $s \leq t$ that

$$P(M_{\frac{1}{\mu} \Lambda_t}^* \mid G_{\frac{1}{\mu} \Lambda_s}) = M_{\frac{1}{\mu} \Lambda_s}^*.$$

Suppose $s \to \Lambda_s$ is strictly increasing. Then certainly $G_{\frac{1}{\mu}\Lambda_s} = F_s$ and the equality reads, because $N^*_{\frac{1}{\mu}\Lambda_t} = N_t$,

$$P(N_t - \Lambda_t | G_t) = N_s - \Lambda_s,$$

which would give Theorem 5.1(a). Of course, to make the proof rigorous, various conditions must be imposed, including the somewhat restrictive one from Proposition 4.6 which bars P from being absorbed. Also, the technique used above in the proof of Theorem 5.1 carries over to the multidimensional case to be studied in Chapter 2, while there the optional sampling theorem has no generally known analogue.

1.6. Statistical models and likelihood ratios.

Formally, a statistical model for counting processes, is a family P of CCP's, and the likelihood function ℓ is the function $\ell(P) = \dfrac{dP}{dP_0}$ on P, where $P_0 \in P$ is a fixed reference probability and $\dfrac{dP}{dP_0}$ is the Radon-Nikodym derivative of P with respect to P_0.

For $\ell(P)$ to make statistical sense it must be assumed that $P \ll P_0$, P is absolutely continuous with respect to P_0. But typically this is not the case: if $P = (\Pi_\mu)_{\mu>0}$ is the family of Poisson probabilities with constant intensities, then $\Pi_\mu \ll \Pi_{\mu_0}$ iff $\mu = \mu_0$ as follows from the observation that $\Pi_\mu(\lim\limits_{t\to\infty} \frac{1}{t} N_t = \mu) = 1$.

What we shall do is therefore to assume that the process is not observed on all of $[0,\infty)$ but only on a finite subinterval $[0,t]$ where $t > 0$. (Alternatively one might consider intervals $[0,\tau]$ where τ is a stopping time such that $P(\tau < \infty) = 1$ for all $P \in P$. Theorem 6.1 carries over verbatim to this situation).

Supposing P,Q to be two CCP's we shall by P^t, Q^t denote the restrictions of the two probabilities to F_t, and then study the derivative $\dfrac{dP^t}{dQ^t}$. For convenience we shall take $Q = \Pi_\mu$, the Poisson probability with constant intensity μ.

It should be remarked that the likelihood function to be given in Theorem 6.1 is mainly useful for statistical inference in parametric models of counting process, and not so vital for the non-parametric models with which we shall later be concerned.

6.1. Theorem. Let P be a stable CCP of class H with locally finite expectations. Then for every $t \geq 0$, $P^t \ll \Pi_\mu^t$ and the Radon-Nikodym derivative is $\ell_t = \dfrac{dP^t}{d\Pi_\mu^t}$ given by

$$(6.2) \qquad \ell_t = \left(e^{-\Lambda_t} \prod_{k=1}^{N_t} \lambda_{\tau_k} - \right) \Big/ \left(\mu^{N_t} e^{-\mu t} \right),$$

where λ_- denotes the intensity and Λ the integrated intensity for P .

Proof. We must show that $P^t(F) = \Pi_\mu(\ell_t;F)$ for all $F \in F_t$. But for this it suffices to consider infinitesimal F of the form

$$F = (\tau_1 \in dt_1, \cdots, \tau_n \in dt_n, N_t = n)$$

which since P and Π_μ are stable, correspond to the atoms of F_t when $n \in \mathbb{N}_0$, $0 < t_1 < \cdots < t_n \le t$. This reduces the problem to studying joint densities for the jump times, which is quite easy. Now

(6.3) $\quad P^t(F) = P(\tau_{n+1} > t | \tau_1 = t_1, \cdots, \tau_n = t_n) \prod\limits_{k=1}^{n} P(\tau_k \in dt_k | \tau_1 = t_1, \cdots, \tau_{k-1} = t_{k-1})$

$$= \exp\left(- \int_{t_n}^{t} ds\ \mu_{nt_1 \cdots t_n}(s)\right)$$

$$\times \prod\limits_{k=1}^{n} \left[\mu_{k-1, t_1 \cdots t_{k-1}}(t_k^-) \exp\left(- \int_{t_{k-1}}^{t_k} ds\ \mu_{k-1, t_1 \cdots t_{k-1}}(s)\right) dt_k\right]$$

$$= e^{-\Lambda_t(\text{on } F)} \prod\limits_{k=1}^{n} \lambda_{\tau_k^-}(\text{on } F)\ dt_k .$$

(Because there is probability 0 of τ_k agreeing with a discontinuity for the function $\mu_{k-1, t_1 \cdots t_{k-1}}$, it is legitimate to write $\mu_{k-1, t_1 \cdots t_{k-1}}(t_k^-)$ above). Similarly, since ℓ_t has to be constant on F

(6.4) $\quad \Pi_\mu^t(\ell_t;F) = \ell_t(\text{on } F) \Pi_\mu^t(F) = \ell_t(\text{on } F)\ \mu^n e^{-\mu t} dt_1 \cdots dt_n ,$

the expression for $\Pi_\mu^t(F)$ being a special case of (6.3) with all $\mu_{kt_1 \cdots t_k} = \mu$. Equating (6.3) and (6.4) and solving for ℓ_t , completes the proof. (Notice that the fact that $P^t \ll \Pi_\mu^t$ really follows because, as is obvious, the P-distribution of $(\tau_1, \cdots, \tau_n)$ is absolutely continuous with respect to the Π_μ-distribution of $(\tau_1, \cdots, \tau_n)$). ∎

On the set $(N_t = 0)$, the product appearing in (6.2) is empty, hence equals 1. Thus

$$\ell_t = \exp\left(-\int_0^t \mu_0 + \mu t\right) \qquad \text{on} \quad (N_t = 0).$$

The next result is given because of its general importance, although we shall not be using it.

6.5. Proposition. If P is stable, the process $\ell = (\ell_t)_{t \geq 0}$ given by (6.2) is a Π_μ-martingale.

Proof. We must show that $\Pi_\mu(\ell_t; F) = \Pi_\mu(\ell_s; F)$ for $s \leq t$, $F \in F_s$. But then also $F \in F_t$ and applying (6.2) at t and at t replaced by s, shows the integrals to equal $P_t(F) = P_s(F)$. |

The converse is also true: if for an arbitrary CCP, P, ℓ_t is defined by (6.2), and it is assumed that ℓ is a Π_μ-martingale, then necessarily P is stable.

As an elaboration of Theorem 6.1 and Proposition 6.5 it may be shown that if P is stable, then considering the full processes P, Π_μ (not restricted to $[0,t]$) one has $P \ll \Pi_\mu$ iff the Π_μ-martingale is uniformly integrable, in which case $\frac{dP}{d\Pi_\mu} = \ell_\infty \overset{D}{=} \lim_{t \to \infty} \ell_t$.

Notes.

The setup used here with canonical processes differs from what is seen
elsewhere, where all processes are defined on an abstract filtered probability space $(\Omega, A, A_t, \mathbb{P})$ satisfying 'the usual conditions', i.e. the
A_t constitute a right-continuous family of sub σ-algebras of A with
all A_t completed with respect to the probability $\mathbb{P}$.

On the canonical path-spaces W and $\overline{W}$, all members F_t of the
filtration are saturated, i.e. any F-measurable union of F_t-atoms,
whether countable or not, is automatically F_t-measurable. We find this
type of measurable structure much nicer than the usual one, especially
because it permits the purely path-algebraic arguments presented in
these lecture notes. Also, since we are dealing with processes with a
very simple structure, it is perfectly possible to develop a theory
without running into the measure theoretical problems that otherwise
necessitate that all σ-algebras be completed a priori.

When working with the canonical spaces and filtrations, all problems concerning the required measurable structure are resolved once
and for all. The cost is a certain amount of inflexibility, that makes
for instance results concerning transformations of one counting process
into another appear a little clumsy.

A standard reference to the theory of counting processes and more
generally, jump processes, is Boel, Varaiya and Wong (1975). They discuss saturation, but most of the time work under 'the usual conditions'.
An earlier important reference is Brémaud (1972). The theory is also
covered in chapter 18 of Liptser and Shiryayev (1977-78). Most relevant
too are Jacod (1975), Brémaud and Jacod (1977).

As a special case of the Doob-Meyer decomposition theorem, Theorem
5.1 may be generalized to counting processes not of class H. A direct
proof along the lines of the proof of Theorem 5.1 is indicated in Exercise 9 below.

Exercises.

1. Find the distribution of the random variable U defined in Proposition 1.1.4, assuming that $\int_0^\infty ds\ \mu(s) < \infty$. ∎

2. Show that if σ_1, σ_2 are stopping times on W or $\overline{W}$, so are $\sigma_1 + \sigma_2$, $\sigma_1 \wedge \sigma_2$, $\sigma_1 \vee \sigma_2$. Show also that if $(\sigma_n)_{n \geq 1}$ is a monotone (increasing or decreasing) sequence of stopping times, then $\sigma = \lim_{n \to \infty} \sigma_n$ is a stopping time. ∎

3. Let τ be a stopping time on W or $\overline{W}$. Show that

$$\widetilde{F}_\tau = \{F \in F\colon F(\tau \leq t) \in F_t \quad \text{for all} \quad t \geq 0\}$$

is a σ-algebra and that $\widetilde{F}_\tau = F_\tau$. ∎

4. The purpose of this exercise is to derive the Kaplan-Meier estimator by some kind of maximum-likelihood reasoning. As in Example 1.2.7, let $X_1, \cdots, X_r$ be i.i.d with common survivor function G, but do not assume that G is continuous, i.e. G is allowed to have atoms. Further let $u_1, \cdots, u_r$ be fixed censoring times so that X_i is censored at u_i. (Each u_i may take the value ∞, corresponding to no censoring).

Suppose now that the values t_i of p of the X_i are observed, while for the remaining X_i it is only recorded that they are censored. Assume also for convenience that the non-censored variables are $X_1, \cdots, X_p$, while $X_{p+1}, \cdots, X_r$ are censored.

Show that, with $g(t) = G(t-) - G(t)$ the probability G attaches to t, the probability of observing $X_i = t_i$ for $i = 1, \cdots, p$ and observing that X_j is censored for $j = p+1, \cdots, r$, at $u_j < \infty$ equals

(1)
$$\prod_{i=1}^{p} g(t_i) \prod_{j=p+1}^{r} G(u_j).$$

The estimator for G is now any survivor function $\hat{G}$ which maximizes this probability.

To carry out the estimation, let $u_{(1)} \leq \cdots \leq u_{(r-p)} < \infty$ be the u_j ordered according to size. Write $I_k = (u_{(k-1)}, u_{(k)}]$, for $k = 1, \cdots, r-p+1$ with $u_{(0)} = 0$, $u_{(r-p+1)} = \infty$. (Some of the I_k may be empty).

Of course (1) is a function of the $g(t_i)$ and $G_k = G(u_{(k)})$, and these variables must satisfy

$$1 \geq G_1 \geq \cdots \geq G_{r-p}$$

(2)

$$\sum_{i:t_i \in I_k} g(t_i) \leq G_{k-1} - G_k$$

for $k = 1, \cdots, r-p+1$.

Show that for a G maximizing (1) there must be equality in (2). Then show that for the G_k fixed (1) is maximized by taking for $t_i \in I_k$,

$$g(t_i) = \frac{n_i}{m_k} (G_{k-1} - G_k),$$

where n_i is the number of X_i observed to take the value t_i and m_k is the number of X_i observed to take values in I_k.

Inserting this expression for $g(t_i)$ in (1), show that (1) is proportional to

$$\left(\prod_{k=1}^{r-p} (G_{k-1} - G_k)^{m_k} G_k^{c_k} \right) G_{r-p}^{m_{r-p+1}}$$

with $G_0 = 1$ and c_k the number of X_j censored at $u_{(k)}$.

Maximize this as a function of G_{r-p+1}, $G_{r-p}, \cdots, G_1$ successively, at each stage inserting the result of the most recent maximalization, and show thereby that (1) is maximized for any $\hat{G}$ satisfying $\hat{G}_0 = 1$ and

$$\hat{G}_k = (1 - \frac{m_k}{R_{k+}}) \, \hat{G}_{k-1} \qquad (1 \le k \le r-p) \, ,$$

$$\frac{\hat{g}(t_i)}{\hat{G}_{k-1}} = \frac{n_i}{R_{k+}} \qquad (1 \le k \le r-p+1, \; t_i \in I_k) \, ,$$

where $R_{k+} = m_k + \cdots + m_{r-p+1} + c_k + \cdots + c_{r-p}$.

Show that the t_i are the only atoms for $\hat{G}$ and discuss to what extent $\hat{G}$ is uniquely determined as a survivor function.

Show finally, that if all the t_i are distinct, the $\hat{G}$ found here agrees with the $\hat{G}$ from Example 1.2.7.

Hint: this amounts to showing that with $\hat{G}$ as above,

$$\frac{\hat{g}(t_i)}{\hat{G}(t_i-)} = \frac{1}{R_{t_i-}}$$

for $i = 1, \cdots, p$, with R_{t_i-} the size of the population at risk immediately <u>before</u> t_i . The denominator R_{k+} above is the size of the risk set immediately <u>after</u> $u_{(k)}$. ▌

5. Show that for a CCP, P, of class H

$$P(N_h \ge 2) = \int_0^h ds \; \mu_0(s) e^{-\int_0^s \mu_0} \left(1 - \exp\left(-\int_s^h du \; \mu_{1,s}(u)\right)\right) \, .$$

Use this to give an example of a process P for which $\lim_{h \downarrow 0} \frac{1}{h} P(N_h \ge 2)$ exists and is > 0 .

Hint: it is enough to take $\mu_0 \equiv 1$, $\mu_{1,s}(u) = \varphi(s)$ a suitable function of s . ▌

6. Show that in order for a CCP , P , of class H to satisfy

$$\lim_{h \downarrow \downarrow 0} \frac{1}{h} P(N_{t+h} - N_t \geq 2 | F_t) = 0$$

a.s. for all $t \geq 0$, it is sufficient that for all $n \geq 1$,
$0 < t_1 < \cdots < t_{n-1} \leq t$ with t less than the termination point
for $G_{n-1, t_1 \cdots t_{n-1}}$ the function

$$\alpha(u) = \sup_{s \in [t, u]} \mu_{n t_1 \cdots t_{n-1} s}(u)$$

defined for $u \geq t$, be locally right integrable at t , i.e.
$\int_t^{t+h} \alpha < \infty$ for some $h > 0$.

Hint: derive and use an expression for $P(N_{t+h} - N_t \geq 2 | F_t)$ si-
milar to the one in Exercise 5 for the case $t = 0$. ▮

7. Discuss why Proposition 1.4.8 (b) cannot be used to give the stability criterion of Example 1.3.4.

Hint: try ! ▮

8. Let P be a CCP of class H with integrated intensity Λ . Show
that for this P , Theorem 1.3.12 may be formulated as follows:
if σ is a stopping time and $w \in (\sigma < \infty)$, then with respect to
the conditional probability $P(\cdot | F_\sigma)(w)$ the distribution of the
process $(N_{t|w})_{t \geq 0}$ defined on W given by

$$N_{t|w}(w') = \begin{cases} 0 & \text{if } t < \sigma(w) \\ N_t(ww') - N_\sigma(w) & \text{if } t \geq \sigma(w) , \end{cases}$$

is the CCP with integrated intensity

$$\Lambda_{t|w}(w') = \begin{cases} 0 & \text{if } t < \sigma(w) \\ \Lambda_t(ww') - \Lambda_\sigma(w) & \text{if } t \geq \sigma(w) . \end{cases}$$

Here and above ww' denotes for any $w' \in W$, the crossed path

$$(ww')(t) = \begin{cases} w(t) & \text{if } t < \sigma(w) \\ w'(t) & \text{if } t \geq \sigma(w) . \end{cases}$$

(In understanding this result, it may be helpful to recall that $P(\cdot | F_\sigma)(w)$ is proper, i.e. as a probability on W it is concentrated on the F_σ-atom containing w).   ▍

9.   Let P be a CCP such that $P N_t < \infty$,   (but do not make any other assumptions). For every n, $t_1 < \cdots < t_n$ define

$$F_{nt_1 \cdots t_n}(t) = 1 - G_{nt_1 \cdots t_n}(t) \qquad (t \geq t_n) .$$

Next define a stochastic process Λ by

$$\Lambda_t = \sum_{k=1}^{N_t} \int_{(\tau_{k-1}, \tau_k]} F_{k-1, \xi_{k-1}}(ds) \, \frac{1}{G_{k-1, \xi_{k-1}}(s-)}$$

$$+ \int_{(\tau_{N_t}, t]} F_{N_t, \xi_{N_t}}(ds) \, \frac{1}{G_{N_t, \xi_{N_t}}(s-)} .$$

Show that Λ is adapted with $\Lambda_0 = 0$ and right-continuous, non-decreasing paths. (In fact Λ is predictable, cf. Definition 3.1.4).

Copying the argument from the proof of Theorem 1.5.1, show that $P\Lambda_t < \infty$ and that $N-\Lambda$ is a P-martingale.

If P is of class H, Λ is of course the integrated intensity for P. A second special case obtains, when P is of class D (see Section 2.5 below), in which case Λ is the accumulated intensity for P.   ▍

2. MULTIVARIATE COUNTING PROCESSES

2.1. Definition and construction of multivariate counting processes.

We shall define and construct multivariate counting processes in a manner similar to the one used in Chapter 1 for the one-dimensional case.

Let E be a finite set. We shall refer to E as the type-set.

1.1. Definition. A counting process with type-set E on a filtered probability space $(\Omega, A, A_t, \mathbb{P})$ is an adapted stochastic process $K = (K_t)_{t \geq 0}$, each $K_t = (K_t^y)_{y \in E}$ taking values in $\mathbb{N}_0^E$, with every component process $(K_t^y)_{t \geq 0}$ a stable (one-dimensional) counting process, and such that no two components jump at the same time. ∎

It should be emphasized that in contrast with Definition 1.2.1, a counting process with type-set E is stable by definition. Notice also that the process $\bar{K} = (\bar{K}_t)_{t \geq 0}$ given by $\bar{K}_t = \sum_{y \in E} K_t^y$ is a one-dimensional counting process.

The self-exciting filtration for a multivariate counting process is defined in complete analogy with the one-dimensional case.

1.2. Definition. The counting process path-space with type-set E is the subset W^E of $(\mathbb{N}_0^E)^{[0,\infty)}$ consisting of those paths $w = (w^y)_{y \in E} \colon [0,\infty) \to \mathbb{N}_0^E$ for which each component w^y belongs to the stable one-dimensional counting process path-space W , and for which $w^y(t-) \neq w^y(t)$ for some y,t implies $w^z(t-) = w^z(t)$ for $z \neq y$. ∎

The space $(\mathbb{N}_0^E)^{[0,\infty)}$ can be identified with the space $\left(\mathbb{N}_0^{[0,\infty)} \right)^E$. Notice that as a subset of the latter, W^E is a genuine subset of the cartesian product $\underset{y \in E}{\chi} W$ because no two components w^y, w^z

of W^E can jump at the same time.

For $t \geq 0$, define $N_t: W^E \to \mathbb{N}_0^E$ by $N_t(w) = w(t)$, write $N_t = (N_t^y)_{y \in E}$ and introduce $\bar{N}_t = \sum_y N_t^y$, $N_\infty = \lim_{t \to \infty} N_t$. Let F denote the σ-algebra of subsets of W^E generated by $(N_t)_{t \geq 0}$, and write F_t for the σ-algebra generated by $(N_s)_{s \leq t}$. Finally, let F_t^y, $\bar{F}_t$ be the σ-algebras generated by $(N_s^y)_{s \leq t}$ and $(\bar{N}_s)_{s \leq t}$ respectively (so that F_t^y, $\bar{F}_t \subset F_t$).

As in the one-dimensional case, F_t consists precisely of those F-measurable sets which are unions of equivalence classes for the e-quivalence relation $\underset{t}{\sim}$ on W^E given by $w \underset{t}{\sim} w'$ iff $w(s) = w'(s)$ for $0 \leq s \leq t$. Also, if $F_{t+} = \bigcap_{\varepsilon > 0} F_{t+\varepsilon}$, then $F_{t+} = F_t$.

1.3. Definition. A canonical counting process with type-set E is a probability on (W^E, F). |

Definitions 1.2 and 1.3 match Definitions 1.2.2 and 1.2.3. Random times, stopping times, pre-τ σ-algebras for τ a random time may be defined as in Definitions 1.2.8 and 1.2.11, just using the path-space (W^E, F) instead of (W, F).

We shall use the abbreviation CCPE for "canonical counting process with type set E".

1.4. Example. Let $(\mu_y)_{y \in E}$ be constants, $\mu_y > 0$. The Poisson process with type-set E and intensities $\mu = (\mu_y)_{y \in E}$ is the restriction to W^E of the product probability $\Pi_\mu = \underset{y \in E}{\otimes} \Pi_{\mu_y}$ on $\underset{y \in E}{\times} W$, Π_{μ_y} denoting the Poisson probability of Example 1.2.4. The definition is legitimate since obviously $\Pi_\mu(W^E) = 1$. Notice that with respect to Π_μ the process N^y is a (non-canonical) Poisson process with intensity μ_y, and $\bar{N}$ is a Poisson process with intensity $\Sigma \mu_y$. |

1.5. Example. As in Example 1.2.6, let $X_1, \ldots, X_r$ be strictly positive random variables. Defining for $i = 1,\ldots,r$

$$K_t^i = 1_{(X_i \leq t)} \, ,$$

the process $K = (K_t)_{t \geq 0}$ with $K_t = (K_t^i)_{1 \leq i \leq r}$ may be viewed as a counting process with type-set $E = \{1,\ldots,r\}$, provided no two X_i are ever equal. Observing the process K is equivalent to observing all X_i. ∎

1.6. Example. When discussing the model for censored survival data of Example 1.2.7, we mentioned that in order to keep track of when censorings occur it may not be enough to count just the number K of observed deaths. So we introduce a second process K^* by

$$K_t^* = \sum_{i=1}^{r} 1_{(U_i \leq t \wedge X_i)} \, .$$

Then K_t^* counts the number of X_i censored before time t and the pair (K,K^*) defines a two-dimensional counting process, provided K and K^* never jump simultaneously (which is a genuine assumption). Observe that the Kaplan-Meier estimator (with random U_i rather than the fixed u_i considered at that stage of Example 1.2.7), is determined by (K,K^*): the population at risk has size $R_{s-} = r - K_{s-} - K_{s-}^* \, .$

The present example is included merely to indicate how more information about the given data may be incorporated in a counting process setup. Later on, when discussing estimation in the model, we shall consider the $U_i = u_i$ as fixed, i.e. work conditionally on the U_i, and assume that the survival times are independent and identically distributed and independent of the U_i. ∎

We shall now discuss the construction of CCPE's.

For $n \geq 1$, write $\tau_n = \inf\{t > 0 : \bar{N}_t = n\}$, the time of the n'th jump of N, and $\tau_n^y = \inf\{t > 0 : N_t^y = n\}$, the time of the n'th jump of the y'th component. Further define $Y_n = y$ on $(N_{\tau_n}^y - N_{\tau_n-}^y = 1)$, so

that Y_n is the component jumping at time τ_n. (Note that Y_n is defined only on $(\tau_n < \infty)$). Then the sequence $(\tau_n)_{n \geq 1}$ completely specifies $\bar{N}$, the sequence (τ_n, Y_n) determines N itself and the sequence
(τ_n^y) describes N^Y.

To study a CCPE, P, we consider for $n \geq 0$ the conditional distribution of τ_{n+1} given $(\tau_1, Y_1, \cdots, \tau_n, Y_n)$ and the conditional distribution of Y_{n+1} given $(\tau_1, Y_1, \cdots, \tau_n, Y_n, \tau_{n+1})$. (For $n = 0$ this
is just the marginal distribution of τ_1 and the conditional distribution of Y_1 given τ_1).

For $n \geq 1$ introduce $\xi_n = (\tau_1, \cdots, \tau_n; Y_1, \cdots, Y_n)$ and define
F_{τ_n} as the σ-algebra generated by $(\xi_{n-1}, \tau_n) = (\tau_1, \cdots, \tau_n; Y_1, \cdots Y_{n-1})$.
Of course F_{τ_n} is the σ-algebra generated by ξ_n. Conditioning on
$F_{\tau_n^-}$ amounts to conditioning on the behavior of the process up to the
time of the n'th jump, but not on the value of that jump.

If we write, for $n \geq 0$

$$G_{n, \xi_n}(t) = P(\tau_{n+1} > t | F_{\tau_n}),$$

$$\pi_{n, \xi_n}(\tau_{n+1}^-, y) = P(Y_{n+1} = y | F_{\tau_{n+1}^-}),$$

then almost surely on the set $(\tau_n < \infty)$, G_{n, ξ_n} is the survivor function
for a probability on $(\tau_n, \infty]$, and almost surely on $(\tau_{n+1} < \infty)$,
$\pi_{n, \xi_n}(\tau_{n+1}^-, \cdot)$ is the density for a probability on E. (For $n = 0$,
we just have functions G_0 and $\pi_0(\tau_1^-, y)$). With this in mind and copying the proof of Theorem 1.3.1 we arrive at the following result.

1.7. **Theorem.** Suppose given for $n \in \mathbb{N}_0$ and any $0 < t_1 < \cdots < t_n$,
$Y_1, \cdots, Y_n \in E$ a probability on the interval $(t_n, \infty]$ with survivor function $G_{n, t_1 \cdots t_n Y_1 \cdots Y_n}$ such that the mapping
$(t_1, \cdots, t_n, Y_1, \cdots, Y_n) \to G_{n, t_1 \cdots t_n Y_1 \cdots Y_n}(t)$ is measurable for all t.
Also suppose given for $n \in \mathbb{N}_0$ and any $0 < t_1 < \cdots < t_{n+1}, Y_1, \cdots, Y_n \in E$,
the density $\pi_{n, t_1 \cdots t_n Y_1 \cdots Y_n}(t_{n+1}^-, \cdot)$ for a probability on E, such

that the mapping $(t_1, \cdots, t_{n+1}, y_1, \cdots, y_n) \to \pi_{n, t_1 \cdots t_n y_1 \cdots y_n}(t_{n+1}^-, y)$
is measurable for all $y \in E$. Then there exists a unique canonical
counting process P with type-set E such that for $n \in \mathbb{N}_0, t > 0, y \in E$

$$(1.8) \qquad P(\tau_{n+1} > t | F_{\tau_n}) = G_{n, \xi_n}(t) \qquad \text{P-a.s.} \quad \text{on} \quad (\tau_n < \infty),$$

$$(1.9) \qquad P(Y_{n+1} = y | F_{\tau_{n+1}^-}) = \pi_{n, \xi_n}(\tau_{n+1}^-, y) \qquad \text{P-a.s.} \quad \text{on} \quad (\tau_{n+1} < \infty),$$

provided, with these specifications of the distributions of jump times and jumps
the time τ_n converges in probability to ∞ as $n \to \infty$. ∎

Remark. The last condition in the theorem guarantees that the constructed P
is stable, as it must be by definition. ∎

The notation $\pi_{n, \xi_n}(\tau_{n+1}^-, y)$ earlier and in (1.9) conforms with the
setup of the next section, where we demand that the conditional proba-
bilities $P(Y_{n+1} = y | F_{\tau_{n+1}^-}, \tau_{n+1} = t)$ may be chosen as left-contin-
uous functions of t.

1.10. Example. For the Poisson process Π_μ of Example 1.4,

$$G_{n t_1 \cdots t_n y_1 \cdots y_n}(t) = \exp(-\bar{\mu}(t - t_n)) \qquad (t \geq t_n),$$

$$\pi_{n t_1 \cdots t_n y_1 \cdots y_n}(t_{n+1}^-, y) = \frac{\mu_y}{\bar{\mu}},$$

writing $\bar{\mu} = \sum_{y \in E} \mu_y$. ∎

The following two examples are important. They show how Markov
chains on a finite state-space may be viewed as multivariate counting
processes.

1.11. Example. Let $(X_t)_{t \geq 0}$ be a Markov chain with stationary transition probabilities on a finite state-space S. Assuming the paths to

be right-continuous with left-limits, the process is completely specified by the sequence $(T_n)_{n \geq 1}$ of jump times and the sequence $(J_n)_{n \geq 0}$
of elements of S with J_n the state reached by the process at the
time of the n'th jump, and J_0 the initial state X_0. The distribution of the sequence $((T_n, J_n))$ is determined by the distribution of
J_0, the initial distribution, and by real-valued parameters
$(\mu_i)_{i \in S}$, $(\pi_{ij})_{i, j \in S, i \neq j}$ with $\mu_i \geq 0$, $\pi_{ij} \geq 0$, $\sum_{j, j \neq i} \pi_{ij} = 1$ such that

(1.12) $\mathbb{P}(T_{n+1} > t | T_1, \cdots, T_n; J_0, \cdots, J_n) = \exp(-\mu_{J_n}(t - T_n))$ $(t \geq T_n)$

on $(T_n < \infty)$ and

(1.13) $\mathbb{P}(J_{n+1} = j | T_1, \cdots, T_{n+1}, J_0, \cdots, J_n) = \pi_{J_n j}$

on $(T_{n+1} < \infty)$.

Let $E = \{(i, j) \in S^2 : i \neq j\}$ and define for $t \geq 0$, $(i, j) \in E$,
$K_{ij}(t)$ as the number of jumps (transitions) from i to j for (X_t)
on the interval $[0, t]$. Then $K = ((K_{ij}(t))_{(i, j) \in E})_{t \geq 0}$ is a counting
process with type-set E, adapted to the self-exciting filtration for
(X_t).

The jumps of K are of course interrelated in the following manner: a jump in component $(i, j) \in E$ must be followed by a jump in one
of the components (j, k). This means that for the CCPE generated by
K, the conditional jump time distributions and conditional jump probabilities need only be determined for strings $y_1, \cdots, y_n$ of types
$y_m = (i_m, j_m) \in E$ for which $j_m = i_{m+1}$ for $1 \leq m \leq n-1$. We shall
call such strings connected.

The counting process K jumps at the same times T_n as the Markov chain X. While knowledge of X on $[0, t]$ determines K on
$[0, t]$, the converse is only true if $t \geq T_1$: to obtain the initial
state X_0 of X, it is necessary to observe K up to time T_1.
Then, if the (i, j)'th component jumps first, $X_0 = i$.

It should now be clear that the CCPE generated by K is given by

$$(1.14) \qquad G_0(t) = \sum_{i \in S} \mathbb{P}(X_0 = i) e^{-\mu_i t} \, ,$$

$$(1.15) \qquad \pi_0(t-,y) = \frac{\mathbb{P}(X_0=i) \mu_i e^{-\mu_i t}}{\sum_{k \in S} \mathbb{P}(X_0=k) \mu_k e^{-\mu_k t}} \, \pi_{ij} \, ,$$

$$(1.16) \qquad G_{n t_1 \cdots t_n y_1 \cdots y_n}(t) = e^{-\mu_{j_n}(t-t_n)} \qquad (n \geq 1, t \geq t_n),$$

$$(1.17) \qquad \pi_{n t_1 \cdots t_n y_1 \cdots y_n}(t-,y) = \pi_{j_n j} = \pi_{ij} \qquad (n \geq 1, t > t_n),$$

writing $y_m = (i_m, j_m)$, $y = (i,j)$ with $y_1, \cdots, y_n$ and $y_1, \cdots, y_n, y$ connected strings.

It is seen that the initial distribution enters only in (1.14), (1.15) and that these two expressions simplify and conform with (1.16), (1.17) if the law of X_0 is degenerate. We shall only consider this situation in the sequel, and write P^i for the CCPE generated by X if $\mathbb{P}(X_0 = i) = 1$. Thus, for P^i

$$G_0(t) = e^{-\mu_i t}, \quad \pi_0(t-,y) = \pi_{ij} \qquad (y = (i,j)),$$

while for $n \geq 1$ the remaining G_n, π_n are given by (1.16), (1.17).

For the statistical analysis of a Markov chain viewed as a CCPE, the idea of considering the initial distribution to be degenerate, a-mounts to conditioning on X_0.

Consider now the family $(P^i)_{i \in S}$ of CCPE's. Writing $Y_n = (I_n, J_n)$ (so that I_n, J_n take values in S), the Markov property of the original process X implies that for $i \in S$, $s, t \geq 0$, $0 \leq m_y \leq n_y$

$$(1.18) \qquad P^i(N^y_{s+t} = n_y, \, y \in E | F_s) = P^{J_{\bar{N}(s)}}(N^y_t = n_y - m_y, \, y \in E)$$

on the set $(N^y_s = m_y, y \in E)$. (We have just used that N^y_{s+t} is N^y_s plus the number of jumps of component y on $(s, s+t]$, together with the fact that the underlying Markov process starts afresh at time s in the state reached at that time, which is $J_{\bar{N}(s)}$).

The identity (1.18) is interesting because it shows that (N_t) is <u>not</u> a Markov process under P^i: $J_{\overline{N}(s)}$ cannot be found from N_s alone, but depends on the entire past $(N_u)_{u \leq s}$. |

1.19. <u>Example</u>. For $(X_t)_{t \geq 0}$ a Markov chain on the finite state-space S with not necessarily stationary transition probabilities, the conditional distributions (1.12), (1.13) of Example 1.11 take the form

$$(1.20) \qquad \mathbb{P}(T_{n+1} > t | T_1, \cdots, T_n, J_1, \cdots, J_n) = \frac{G^{(J_n)}(t)}{G^{(J_n)}(T_n)} \qquad (t \geq T_n),$$

$$\mathbb{P}(J_{n+1} = j | T_1, \cdots, T_{n+1}, J_1, \cdots, J_n) = \pi_{J_n j}(T_{n+1}-)$$

where for every $i \in S$, $G^{(i)}$ is a survivor function for a probability on $(0, \infty]$, and for every $t > 0$, $\pi_{ij}(t-) \geq 0$, $\sum_{j, j \neq i} \pi_{ij}(t-) = 1$ and the functions $t \to \pi_{ij}(t-)$ are measurable for all $(i,j) \in E$.

To avoid technicalities we shall assume that all the $G^{(i)}$ have termination point ∞, so that the denominator on the right of (1.20) is always strictly positive. Also then the Markov chain will with probability one only have finitely many jumps on any finite time-interval.

With these conditions, the conditional probabilities above do not describe all Markov chains on a finite state-space with right-continuous, left-limit paths and finitely many jumps on finite intervals. We shall return to this problem, when discussing the interpretation of some estimators of the $G^{(i)}$ in Section 4.4.4.

Defining K and K_{ij} as in Example 1.11 K becomes a counting process. If $\mathbb{P}(X_0 = i) = 1$, the CCPE P^i generated by K is given by

$$G_0(t) = G^{(i)}(t), \quad \pi_0(t-, y) = \pi_{ij}(t-),$$

$$G_{nt_1 \cdots t_n y_1 \cdots y_n}(t) = \frac{G^{(j_n)}(t)}{G^{(j_n)}(t_n)} \qquad (n \geq 1, t \geq t_n),$$

$$\pi_{nt_1 \cdots t_n y_1 \cdots y_n}(t-, y) = \pi_{j_n j}(t-) = \pi_{ij}(t-) \quad (n \geq 1, t > t_n)$$

for connected strings $y_1, \cdots, y_n$ and $y_1, \cdots, y_n, y$ with $y_m = (i_m, j_m)$, $y = (i,j)$.

1.21. **Example.** Let $X_1, \cdots, X_r$ be i.i.d with survivor function G. The process $K = (K_t)_{t \geq 0}$ where $K_t = \sum_{i=1}^{r} 1_{(X_i \leq t)}$ will be a counting process if G has no atoms (see Example 1.2.6 and Example 1.3.7), but not otherwise. If G has atoms one may instead consider the r-dimen - sional process $(K^1, \cdots, K^r)$ where K_t^y is the number of jumps of size y for K on the interval $[0,t]$, i.e. $K_t^y = \sum_{s \leq t} 1_{(K_s - K_{s-} = y)}$.

For the CCPE on $E = \{1, \cdots, r\}$ generated by $(K^1, \cdots, K^r)$ one finds, arguing as in Example 1.3.8 that

$$G_{nt_1 \cdots t_n y_1 \cdots y_n}(t) = \left(\frac{G(t)}{G(t_n)} \right)^{r - \bar{y}} \qquad (t \geq t_n),$$

$$\pi_{nt_1 \cdots t_n y_1 \cdots y_n}(t-, y) = \begin{cases} 1 & (t > t_n, g(t) = 0, y = 1) \\ p & (t > t_n, g(t) > 0, y \geq 1), \end{cases}$$

where

$$p = \binom{r - \bar{y}}{y} p_t^y (1 - p_t)^{r - \bar{y} - y} / [1 - (1 - p_t)^{r - \bar{y}}],$$

writing $\bar{y} = \Sigma y_i$, $p_t = g(t)/G(t-)$ with $g(t) = G(t-) - G(t)$ the proba- bility mass G gives to the point t.

We shall conclude this   section with the multidimensional analogue of Theorem 1.3.12. So let P be a CCPE with conditional jump time dis- tributions and conditional jump probabilities as in (1.8), (1.9).

For σ a given random time, let for $n \geq 1$, $\tau_{\sigma,n}$ be the time of the n'th jump of N after σ, and let $Y_{\sigma,n}$ denote the component of N jumping at time $\tau_{\sigma,n}$.

The sequence $(\tau_{\sigma,1}, Y_{\sigma,1}, \tau_{\sigma,2}, Y_{\sigma,2}, \cdots)$ describes the counting process N^* where $N_u^* = N_u - N_\sigma$, and we shall study N^* with respect to the conditional probability $P(\cdot|F_\sigma)$.

For notation we shall write $P_w^{F_\sigma}$ instead of $P(\cdot|F_\sigma)(w)$. Also, we shall use $(\xi_{\overline{N}_\sigma}(w), \xi_{\sigma,n})$ as shorthand for $(\tau_1(w), \cdots, \tau_{\overline{N}_\sigma}(w), \tau_{\sigma,1}, \cdots, \tau_{\sigma,n}, Y_1(w), \cdots, Y_{\overline{N}_\sigma}(w), Y_{\sigma,1}, \cdots, Y_{\sigma,n})$. The next result is stated without proof but may be argued exactly as Theorem 1.3.12.

1.22. Theorem. For every stopping time σ and every $w \in W^E$

$$P_w^{F_\sigma}(\tau_{\sigma,n+1} > u | \tau_{\sigma,1}, \cdots, \tau_{\sigma,n}, Y_{\sigma,1}, \cdots, Y_{\sigma,n}) = \begin{cases} \dfrac{G_{\overline{N}_\sigma(w), \xi_{\overline{N}_\sigma}(w)}(u)}{G_{\overline{N}_\sigma(w), \xi_{\overline{N}_\sigma}(w)}(\sigma)} & (u \geq \sigma,\ n = 0) \\ \\ G_{\overline{N}_\sigma(w)+n, \xi_{\overline{N}_\sigma}(w), \xi_{\sigma,n}}(u) & (u \geq \tau_{\sigma,n}, n \geq 1), \end{cases}$$

$$P_w^{F_\sigma}(Y_{\sigma,n+1} = y | \tau_{\sigma,1}, \cdots, \tau_{\sigma,n+1}, Y_{\sigma,1}, \cdots, Y_{\sigma,n}) = \pi_{\overline{N}_\sigma(w)+n, \xi_{\overline{N}_\sigma}(w), \xi_{\sigma,n}}(\tau_{\sigma,n+1}, y)$$

$$(y \in E, \quad n \geq 0). \quad \blacksquare$$

2.2. Intensities and martingale representations.

Consider the Poisson process Π_μ from Examples 1.4 and 1.10. Comparing with the one-dimensional case, it is natural that the intensity for Π_μ should be the vector $(\mu_y)_{y \in E}$. Then of course $\Sigma \mu_y$ is the intensity for $\bar{N}$ under Π_μ.

We shall now first define the right-continuous regularization of the intensity process for a suitable class of CCPE's and show that it is a limit of conditional probabilities, just as in the one-dimensional case (see Proposition 1.4.2). Then we shall proceed to define the intensity process itself.

Suppose that P is a CCPE with all $G_{nt_1 \ldots t_n y_1 \ldots y_n}$ having smooth densities with intensity functions $\mu_{nt_1 \ldots t_n y_1 \ldots y_n}$, and such that the mapping $t \to \pi_{nt_1 \ldots t_n y_1 \ldots y_n}(t-,y)$ is left-continuous with right-limits everywhere on (t_n, ∞) and $\lim_{t \downarrow\downarrow t_n} \pi_{nt_1 \ldots t_n y_1 \ldots y_n}(t-,y)$ exists. We shall denote the class of all such P by H^E.

Suppose that P is of class H^E, and write

$$\pi_{nt_1 \ldots t_n y_1 \ldots y_n}(t,y) = \lim_{s \downarrow\downarrow t} \pi_{nt_1 \ldots t_n y_1 \ldots y_n}(s-,y).$$

Now consider the process $\lambda = (\lambda_t)_{t \geq 0}$ on (W^E, F) taking values in $[0,\infty]^E$ given by $\lambda_t = (\lambda_t^y)_{y \in E}$ with

(2.1) $\lambda_t^y = \mu_{\bar{N}_t, \xi_{\bar{N}_t}}(t) \pi_{\bar{N}_t, \xi_{\bar{N}_t}}(t,y)$ $(t \geq 0, y \in E)$.

Also introduce $\bar{\lambda} = \sum_y \lambda^y$.

2.2. **Proposition.** The process λ is F_t-adapted and has sample paths which are P-almost surely right-continuous with left-limits everywhere, and locally integrable in the sense that for P-almost all $w \in W^E$, $\int_0^t ds \bar{\lambda}_s(w) < \infty$ for all $t \geq 0$. Finally, for $t \geq 0$, $y \in E$, P-almost surely

(2.3) $\lim_{h \downarrow\downarrow 0} \frac{1}{h} P(\bar{N}_{t+h} - \bar{N}_t \geq 1, Y_{t,1} = y \mid F_t) = \lambda_t^y$,

(2.4) $$\lim_{h \downarrow\downarrow 0} \frac{1}{h} P(\overline{N}_{t+h} - \overline{N}_t \geq 1 | F_t) = \overline{\lambda}_t .$$

Remark. In comparison with Proposition 1.4.2, the statement here is stronger: "right local integrability" is replaced by the more restrictive "local integrability". The reason is that multivariate counting processes are stable by definition, while the one-dimensional ones are not.

Proof. The right-continuity, left-limit properties are obvious. The local integrability follows because P is stable, adapting the proof of Proposition 1.4.8 (b) to the multivariate situation. Since (2.4) follows from (2.3) summing on y, it only remains to prove (2.3). But using Theorem 1.22 (with $\sigma = t$) it is seen that

$$P(\overline{N}_{t+h} - \overline{N}_t \geq 1, Y_{t,1} = y | F_t) = P(\tau_{t,1} \leq t+h, Y_{t,1} = y | F_t)$$

$$= \int_t^{t+h} ds\ \mu_{\overline{N}_t, \xi_{\overline{N}_t}}(s) \exp(-\int_t^s \mu_{\overline{N}_t, \xi_{\overline{N}_t}})\ \pi_{\overline{N}_t, \xi_{\overline{N}_t}}(s-, y) .$$

Now divide by h and let $h \downarrow\downarrow 0$. ∎

In nice cases an alternative description of the intensity is available.

2.5. Proposition. The process λ satisfies

$$\lim_{h \downarrow\downarrow 0} \frac{1}{h} P(N_{t+h}^y - N_t^y \geq 1 | F_t) = \lambda_t^y$$

for all $t \geq 0$ such that

(2.6) $$\lim_{h \downarrow\downarrow 0} \frac{1}{h} P(\overline{N}_{t+h} - \overline{N}_t \geq 2 | F_t) = 0.$$

Proof. If (2.6) holds,

$$\lim_{h \downarrow\downarrow 0} \frac{1}{h} [P(N_{t+h}^y - N_t^y \geq 1 | F_t) - P(N_{t+h}^y - N_t^y \geq 1, \overline{N}_{t+h} - \overline{N}_t = 1 | F_t)] = 0 ,$$

and here the last probability equals $P(\overline{N}_{t+h} - \overline{N}_t = 1, Y_{t,1} = y | F_t)$.

A second application of (2.6) combined with Proposition 2.2 now gives
the desired result. |

Recall that (2.6) is not always satisfied, cf. Section 1.4.

As in the one-dimensional case it is found that although the right-
continuous process λ is not uniquely determined from P, any two
versions are indistinguishable. We may therefore define the intensity
itself as in the one-dimensional case.

2.7. <u>Definition</u>. For a canonical counting process P of class H^E,
the <u>intensity process</u> $\lambda_- = (\lambda_{t-})_{t>0}$ with $\lambda_{t-} = (\lambda_{t-}^y)$, is given by
$\lambda_{t-}^y = \lim_{s \uparrow\uparrow t} \lambda_s^y$. |

Thus the intensity is defined almost surely, left-continuous on
$(0,\infty)$ with right-limits on $[0,\infty)$ and locally integrable. In terms
of the μ and π, we have (except on a null set)

$$(2.8) \quad \lambda_{t-}^y = \mu_{\overline{N}_{t-},\, \xi_{\overline{N}_{t-}}}(t-) \, \pi_{\overline{N}_{t-},\, \xi_{\overline{N}_{t-}}}(t-,y) \qquad (t > 0, \ y \in E).$$

2.9. <u>Proposition</u>. Suppose $\lambda_- = (\lambda_{t-})_{t>0}$, where $\lambda_{t-} = (\lambda_{t-}^y)_{y \in E}$,
is a $[0,\infty]^E$-valued F_t-adapted process with sample paths all of
which are left-continuous on $(0,\infty)$ with right-limits on $[0,\infty)$.
Suppose also that $\int_0^t ds \overline{\lambda}_s(w) < \infty$ for all $w \in W^E$, $t \geq 0$. Then there
is a unique CCPE of class H^E with λ_- as intensity process.

<u>Proof</u>. Given λ_-, define $\lambda = (\lambda^y)_{y \in E}$ by $\lambda_t^y = \lim_{s \downarrow\downarrow t} \lambda_{s-}^y$. Arguing
as in the proof of Proposition 1.4.4 with (2.8) in mind, an obvious
candidate for the P we are looking for, should satisfy

$$\mu_{nt_1 \ldots t_n y_1 \ldots y_n}(t-) = \overline{\lambda}_{t-}(w), \quad \pi_{nt_1 \ldots t_n y_1 \ldots y_n}(t-,y) = \frac{\lambda_{t-}^y(w)}{\overline{\lambda}_{t-}(w)}$$

for any (hence all) w with $\overline{N}_{t-}(w) = n$, $\tau_1(w) = t_1, \ldots,$
$\tau_n(w) = t_n$, $Y_1(w) = y_1, \ldots, Y_n(w) = y_n$. There are now two problems

in actually constructing P. Firstly, the μ and π must satisfy some conditions in order that P be stable, cf. the last condition of Theorem 1.7. Secondly, the definition of $\pi_{nt_1 \ldots t_n y_1 \ldots y_n}$ above leaves that quantity unspecified if $\overline{\lambda}_{t-}(w) = 0$. The first problem is taken care of by the assumption that $\int_0^t ds \overline{\lambda}_s < \infty$ together with Proposition 1.4.8 (b) (adapted to the multivariate situation). The second is resolved like this: from the expression for $\mu_{nt_1 \ldots t_n y_1 \ldots y_n} = \mu$ in terms of $\overline{\lambda}$ it is seen that the set of t where $\pi_{nt_1 \ldots t_n y_1 \ldots y_n}(t-,y) = \pi$ is not defined is $A = \{t > t_n : \mu(t-) = 0\}$. But on that set the definition of π does not matter because

$$P(\tau_{n+1} \in A | \tau_1 = t_1, \ldots, \tau_n = t_n, \, Y_1 = y_1, \ldots, Y_n = y_n) = \int_A ds \mu(s) e^{-\int_{t_n}^s \mu} = 0,$$

so if the first n jumps occur at $t_1, \ldots, t_n$, the n+1'st jump will never occur at a time point in A.

That the P constructed this way is unique, is argued as in the proof of Proposition 1.4.4. ∎

With λ the intensity, we write $\Lambda = (\Lambda^y)_{y \in E}$ for the integrated intensity, i.e. $\Lambda_t^y = \int_0^t ds \, \lambda_s^y$, and also introduce $\overline{\Lambda}_t = \int_0^t ds \overline{\lambda}_s = \sum_y \Lambda_t^y$.

In analogy with Proposition 1.4.8, the next result summarizes some properties of λ_- and Λ. Remember that we are now only considering stable processes.

2.10. <u>Proposition</u>. Let P be a CCPE of class H^E. Then

(a)   For all $n \in \mathbb{N}_0$, $\overline{\Lambda}_{\tau_n} < \infty$ P-a.s., $\overline{\Lambda}_{\tau_n} < \overline{\Lambda}_{\tau_{n+1}}$ P-a.s. on

$(\tau_{n+1} < \infty)$ and $\Lambda^y(\tau_n^y) < \Lambda^y(\tau_{n+1}^y)$ P-a.s. on $(\tau_{n+1}^y < \infty)$ for

all $y \in E$.

(b)   One has $\overline{\Lambda}_\infty = \infty$ P-a.s. on $(\overline{N}_\infty = \infty)$.

(c)   For all $n \in \mathbb{N}_0$, $y \in E$, $\lambda^y(\tau_n^y-) > 0$ P-a.s. on $(\tau_n^y < \infty)$.

<u>Proof</u>. The proofs of Proposition 1.4.8, slightly modified, carry over
to this new case. As an illustration we show that $\Lambda^y(\tau_n^y) < \Lambda^y(\tau_{n+1}^y)$
P-a.s. on $(\tau_{n+1}^y < \infty)$. But for this it is enough to show that for all
$n \geq 0$, $\Lambda_{\tau_n}^y < \Lambda_{\tau_{n+1}}^y$ P-a.s. on $(\tau_{n+1} < \infty, Y_{n+1} = y)$. We find that

$$P(\Lambda_{\tau_n}^y = \Lambda_{\tau_{n+1}}^y, \ Y_{n+1} = y, \ \tau_{n+1} < \infty \mid F_{\tau_n})$$

$$= \int_{\tau_n}^{\infty} ds \ \mu(s) \exp\left(-\int_{\tau_n}^{s} \mu\right) \pi(s) \ 1_{\left(\int_{\tau_n}^{s} \mu\pi = 0\right)} (s) = 0 ,$$

writing $\mu(s) = \mu_{n\xi_n}(s)$, $\pi(s) = \pi_{n\xi_n}(s-,y)$.   ▌

We shall conclude this part of the section with the multidimensional version of the time substitution result 1.4.6. The proof is much more
difficult than in the one-dimensional case, and is only skeleted. It can
safely be omitted!

To formulate the result, introduce the inverses

$$\rho_u^y(w) = \inf\{t > 0 : \Lambda_t^y(w) > u\} ,$$

and observe that since $(\rho_u^y \geq t) = (\Lambda_t^y \leq u)$, each ρ_u^y is a stopping
time.

2.11. <u>Proposition</u>. Suppose that P is a CCPE of class H^E with integrated intensity $(\Lambda_t)_{t \geq 0}$, and let $\mu = (\mu_y)_{y \in E}$ be a vector of con-

stants, $\mu_y > 0$. Then, if $P(N_\infty^y = \infty) = 1$ for all $y \in E$, the stochastic process N^* defined by

$$N_u^{*y} = N_{\rho^y(\mu_y u)}^y \qquad (u \geq 0, y \in E)$$

is a Poisson process with intensity μ.

<u>Sketch of proof</u>. The proof is made difficult by the fact that each component is submitted to its own timesubstitution, so that for the new process N^* the jumps of the different components occur in a totally different order from those of N. This rules out a proof similar to that of Proposition 1.4.6. Instead, what one does is to use the same timesubstitution to all components simultaneously. This is then combined with an induction argument in k, where k is the cardinality of E.

So fix $y_0 \in E$ and define $\tilde{N}$ by

$$\tilde{N}_u^y = N_{\rho^{y_0}(\mu_{y_0} u)}^y \qquad (u \geq 0, y \in E).$$

Here a new complication arises since $\tilde{N}$ need not be a counting process: since $t \to \Lambda_t^{y_0}$ may be flat on some intervals, parts of the original process can be left out by the time substitution, and although component y_0 will never jump on the left out pieces (cf. Proposition 2.10(a)), the remaining components may well have jumps there.

We shall therefore assume that $t \to \Lambda_t^{y_0}$ is strictly increasing. By the assumption that $P(N_\infty^{y_0} = \infty) = 1$ and an unproved sharpening of Proposition 2.10(b), $\Lambda_t^{y_0} \uparrow \infty$ P-a.s. as $t \uparrow \infty$, so that $u \to \rho_u^{y_0}$ is strictly increasing and continuous on $[0,\infty)$ with $\rho_0^y = 0$. Consequently $\tilde{N}$ is a stable counting process and we denote by $\tilde{P}$ the CCPE generated by $\tilde{N}$.

The next step consists in finding the intensity $\tilde{\lambda}$ for $\tilde{P}$. If $T^{y_0}: W^E \to W^E$ is defined by $N_u \circ T^{y_0} = N_{\rho^{y_0}(\mu_{y_0} u)}$ (so that $\tilde{P} = T^{y_0}P$), one finds by sheer calculation that

$$\tilde{\lambda}_u^y \circ T^{y_0} = \frac{\lambda^y_{\rho} y_0 (\mu_{y_0} u)}{\lambda^{y_0}_{\rho} y_0 (\mu_{y_0} u)} \; \mu_{y_0} \; ,$$

in particular $\tilde{\lambda}_u^{y_0} \circ T^{y_0} = \mu_{y_0}$. The integrated intensity for $\tilde{P}$ therefore satisfies

(2.12) $$\tilde{\Lambda}_u^y \circ T^{y_0} = \Lambda^y_{\rho} y_0 (\mu_{y_0} u) .$$

The structure of $\tilde{\lambda}$ ensures that marginally under P, the process $\tilde{N}^{y_0}$ is Poisson with intensity μ_{y_0}, and furthermore that for every u, $(\tilde{N}_v^{y_0} - \tilde{N}_u^{y_0})_{v \geq u}$ and $(\tilde{N}_v)_{v \leq u}$ are stochastically independent. (But it is not true that $\tilde{N}^{y_0}$ is independent of $(\tilde{N}^y)_{y \neq y_0}$.)

The idea is now under P to study the conditional process $(\tilde{N}^y)_{y \neq y_0}$ given $\tilde{N}^{y_0}$. For each value of the conditioning process this gives a counting process with type-set $E \smallsetminus \{y_0\}$. Because of the independence properties just mentioned, the integrated intensity for the CCP $(E \smallsetminus \{y_0\})$ generated by this process is given by (2.12) (with everything depending on $\tilde{N}^{y_0}$ considered fixed).

If the proposition has been proved for processes with type-sets of cardinality $k-1$, that result may now be used to transform the process $(\tilde{N}^y)_{y \neq y_0}$ given $\tilde{N}^{y_0}$ into a Poisson process $(\tilde{\tilde{N}}^y)_{y \neq y_0}$ with intensity $(\mu_y)_{y \neq y_0}$, (but of course there is a new transformation for each value of the conditioning process. In all however, $(\tilde{\tilde{N}}^y)_{y \neq y_0}$ is just a transformation of the original process N (under P)).

Since obviously $\tilde{N}^{y_0}$ and $(\tilde{\tilde{N}}^y)_{y \neq y_0}$ are independent, the proposition now follows for type-sets of cardinality k, by showing by a straightforward but tedious calculation that $N^{*y_0} = \tilde{N}^{y_0}$, $N^{*y} = \tilde{\tilde{N}}^y$ for $y \neq y_0$. ∎

We conclude this subsection with the multidimensional analogue of Theorem 1.5.1. We shall say that a CCPE, P, has <u>finite expectations lo-</u>

<u>cally</u> if $P\bar{N}_t < \infty$ for all $t \geq 0$.

2.13. <u>Theorem</u>. Suppose P is a CCPE of class H^E with finite expectations locally and define $M = (M^y)_{y \in E}$ by $M_t^y = N_t^y - \Lambda_t^y$, where Λ is the integrated intensity of P. Then with respect to P

(a) $(M_t^y, F_t)_{t \geq 0}$ is a martingale for every $y \in E$;

(b) $(M_t^{y^2} - \Lambda_t^y, F_t)_{t \geq 0}$ is a martingale for every $y \in E$;

(c) $(M_t^y M_t^z, F_t)_{t \geq 0}$ is a martingale for every $y \neq z \in E$.

<u>Proof</u>. In complete analogy with the proof of Theorem 1.5.1, it is shown using Theorem 1.22, that the theorem follows if one proves that

$$PN_{t \wedge \tau_1}^y = P\Lambda_{t \wedge \tau_1}^y = PM_{t \wedge \tau_1}^{y^2} \cdot,$$

(2.14) $$PM_{t \wedge \tau_1}^y M_{t \wedge \tau_1}^z = 0$$

for all $t \geq 0$, $y \neq z \in E$ and all P of class H^E with locally finite expectations. Just as in the one-dimensional case this is only a matter of calculation, so here we shall just prove (2.14). But

$$PM_{t \wedge \tau_1}^y M_{t \wedge \tau_1}^z = P(M_{t \wedge \tau_1}^y M_{t \wedge \tau_1}^z ; \tau_1 > t)$$

$$+ P(M_{t \wedge \tau_1}^y M_{t \wedge \tau_1}^z ; Y_1 \neq y, z, \tau_1 \leq t)$$

$$+ P(M_{t \wedge \tau_1}^y M_{t \wedge \tau_1}^z ; Y_1 = y, \tau_1 \leq t)$$

$$+ P(M_{t \wedge \tau_1}^y M_{t \wedge \tau_1}^z ; Y_1 = z, \tau_1 \leq t)$$

$$= \left(\int_0^t \mu_0 \pi_0^y \right) \left(\int_0^t \mu_0 \pi_0^z \right) e^{-\int_0^t \mu_0}$$

$$+ \int_0^t ds \mu_0(s) e^{-\int_0^s \mu_0} (1 - \pi_0(s-,y) - \pi_0(s-,z)) \left(\int_0^s \mu_0 \pi_0^y \right) \left(\int_0^s \mu_0 \pi_0^z \right)$$

$$- \int_0^t ds \mu_0(s) e^{-\int_0^s \mu_0} \pi_0(s-,y) \left(1 - \int_0^s \mu_0 \pi_0^y \right) \left(\int_0^s \mu_0 \pi_0^z \right)$$

$$- \int_0^t ds \mu_0(s) e^{-\int_0^s \mu_0} \pi_0(s-,z) \left(\int_0^s \mu_0 \pi_0{}^y \right) \left(1 - \int_0^s \mu_0 \pi_0{}^z \right)$$

writing π_0^y as short for the function $s \to \pi_0(s-,y)$. It is readily checked that the right derivative of this is 0 for all t. ∎

Assertion (c) of the theorem shows that for $y \neq z$, the martingales M^y and M^z have conditionally uncorrelated increments: for $s < t$ the conditional covariance given F_s between the increments of the two martingales over $(s,t]$ is

$$P[(M_t^y - M_s^y - P(M_t^y - M_s^y | F_s)) \ (M_t^z - M_s^z - P(M_t^z - M_s^z | F_s)) \ | \ F_s]$$

$$= P((M_t^y - M_s^y)(M_t^z - M_s^z) | F_s) - P(M_t^y - M_s^y | F_s) \ P(M_t^z - M_s^z | F_s).$$

Here the last term disappears because M^y, M^z are martingales, while for the same reason, the first term reduces to

$$P(M_t^y M_t^z | F_s) - M_s^y M_s^z$$

which is 0 because of (c).

This observation will be important later: when developing a theory for the asymptotic distribution of the estimators we shall eventually define, the property of conditionally uncorrelated increments will translate into true stochastic independence between Gaussian processes with independent increments.

In the general theory of processes, the fact that the product $M^y M^z$ is a martingale, is expressed by saying that the martingales M^y and M^z for $y \neq z$ are orthogonal. We shall return to this concept in Chapter 3.

2.3. Products of canonical counting processes.

In the statistical theory we shall consider several independent counting processes at a time. In this section we shall now see, how such a family of processes may be viewed as just one counting process.

Let for $i = 1, \cdots, r$, P_i be a canonical counting process with type-set E_i, and assume that P_i is of class H^{E_i}. Considering the product probability $P = P_1 \otimes \cdots \otimes P_N$ on the product space $\underset{i=1}{\overset{r}{X}} W^{E_i}$, it is clear that with respect to P no two components from arbitrary two P_i jump at the same time, and hence P may be viewed as a canonical counting process with type-set $E^* = \underset{i}{\cup} \{i\} \times E_i$ which is the disjoint union of the E_i.

A path $w \in W^{E^*}$ is of the form $w = (w^1, \cdots, w^r)$, where for each i, $w^i = (w^{i, y_i})_{y_i \in E_i}$ is a path in W^{E_i}. We write $N = (N^{i, y_i})_{1 \le i \le r, y_i \in E_i}$ and write $\tau_1 < \tau_2 < \cdots$ for the sequence of jump times for N, Y_n for the component jumping at τ_n, so Y_n is E^*-valued. We call $N^i = (N^{i, y_i})_{y_i \in E_i}$ the i'th subprocess. Finally we write $\lambda_-^i = (\lambda_-^{i, y_i})_{y_i \in E_i}$ for the intensity process for P_i and $\bar{\lambda}_-^i = \underset{y_i}{\Sigma} \lambda_-^{i, y_i}$.

3.1. Theorem. The product process P is of class H^{E^*}, and its intensity process $\lambda_- = (\lambda_-^{(i, y_i)})_{1 \le i \le r, y_i \in E_i}$ is given by

$$\lambda_{t-}^{(i, y_i)}(w) = \lambda_{t-}^{i, y_i}(w^i).$$

Proof. It is an easy matter to check that $P \in H^{E^*}$. The intensity is most easily found using Proposition 2.2 according to which

$$\lim_{h \downarrow 0} \frac{1}{h} P(\bar{N}_{t+h} - \bar{N}_t \ge 1, Y_{t,1} = (i, y_i) \mid F_t) = \lambda_t^{(i, y_i)}$$

with $\bar{N} = \underset{i, y_i}{\Sigma} N^{i, y_i}$. But for a given w and $h > 0$ so small that w does not jump on $(t, t+h]$ we find, writing $\tau_{t,1}^i$ for the time of the first jump after t of the i'th subprocess, $Y_{t,1}^i$ for the component jumping at $\tau_{t,1}^i$,

$$P(\bar{N}_{t+h} - \bar{N}_t \geq 1, \ Y_{t,1} = (i,y_i) \,|\, F_t)(w)$$

$$= \ P(\tau_{t,1}^i \leq t+h, \ Y_{t,1}^i = y_i, \tau_{t,1}^j > \tau_{t,1}^i, \ j \neq i \,|\, F_t)(w)$$

$$= \int_t^{t+h} ds \ \lambda_s^{i,y_i}(w^i) \exp\left(-\int_t^s du \ \bar{\lambda}_u^i(w^i)\right) \prod_{j \neq i} \exp\left(-\int_t^s du \ \bar{\lambda}_u^j(w^j)\right) \ ,$$

using the independence of the subprocesses, the definition of each λ^{i,y_i} (see (2.1)) and the condition on h assumed above. Now divide by h and let $h \downarrow 0$. ∎

2.4. Likelihood ratios.

Let P be a CCPE and P^t be the restriction to F_t of P. We shall discuss the Radon-Nikodym derivative of P^t with respect to a given Poisson probability Π_μ^t with intensity $\mu = (\mu_y)_{y \in E}$.

4.1 **Theorem.** Let P be a CCPE of class H^E. Then for every $t \geq 0$, $P^t \ll \Pi_\mu^t$ and the Radon-Nikodym derivative is $\ell_t = \dfrac{dP^t}{d\Pi_\mu^t}$ given by

$$\ell_t = \left(e^{-\bar{\Lambda}_t} \prod_{k=1}^{\bar{N}_t} \lambda_{\tau_k-}^{Y_k} \right) \Big/ \left(e^{-\bar{\mu}^t} \prod_{y \in E} \mu_y^{N_t^y} \right)$$

where $\bar{\mu} = \sum_y \mu_y$.

Proof. Proceeding as in the proof of Theorem 1.6.1. we shall show that $P^t(F) = \int_F d\Pi_\mu^t \, \ell_t$ for F an infinitesimal approximation to an arbitrary F_t-atom, i.e.

$$F = (\tau_1 \in dt_1, \cdots, \tau_n \in dt_n, \ Y_1 = y_1, \cdots, Y_n = y_n, \ \bar{N}_t = n)$$

for some $n \in \mathbb{N}_0$, $t_1 < \cdots < t_n \leq t$, $y_1, \cdots, y_n \in E$. But writing $\mu_k(s) = \mu_{k,t_1 \cdots t_k \, y_1 \cdots y_k}(s)$, $\pi_k(s-,y) = \pi_{k,t_1 \cdots t_k \, y_1 \cdots y_k}(s-,y)$, we get

$$P^t(F) = P(\tau_{n+1} > t \mid \tau_1 = t_1, \cdots, \tau_n = t_n, \ Y_1 = y_1, \cdots, Y_n = y_n)$$

$$\prod_{k=1}^{n} \Big[P(\tau_k \in dt_k \mid \tau_1 = t_1, \cdots, \tau_{k-1} = t_{k-1}, \ Y_1 = y_1, \cdots, Y_{k-1} = y_{k-1})$$

$$P(Y_k = y_k \mid \tau_1 = t_1, \cdots, \tau_k = t_k, Y_1 = y_1, \cdots, Y_{k-1} = y_{k-1}) \Big]$$

$$= \exp\left(-\int_{t_n}^{t} ds \mu_n(s) \right) \prod_{k=1}^{n} \mu_{k-1}(t_k) \exp\left(-\int_{t_{k-1}}^{t_k} \mu_{k-1} \right) dt_k \, \pi_{k-1}(t_k-, y_k)$$

$$= \exp\left(-\bar{\Lambda}_t(\text{on } F) \right) \prod_{k=1}^{\bar{N}_t(\text{on } F)} \lambda_{\tau_k-}^{Y_k}(\text{on } F) \, dt_1 \cdots dt_k,$$

while

$$\int_F d\Pi_\mu^t \, \ell_t = \ell_t (\text{on } F) \, e^{-\bar{\mu}t} \prod_{y \in E} \mu_y^{N_t^y (\text{on } F)} \, dt_1 \cdots dt_k. \ \blacksquare$$

It is easily seen that $(\ell_t, F_t)_{t \geq 0}$ is a Π_μ-martingale (cf. Proposition 1.6.5) with paths that are everywhere right-continuous with left-limits.

As in Section 1.6 it can be shown that if P is of class H^E and τ is a stopping time such that $P(\tau < \infty) = \Pi_\mu(\tau < \infty) = 1$, then $P^\tau \ll \Pi_\mu^\tau$, where P^τ, Π_μ^τ are the restrictions to F_τ of P, Π_μ, and

$$\frac{dP^\tau}{d\Pi_\mu^\tau} = \ell_\tau \; .$$

2.5. Discrete counting processes.

We have specified any CCPE, P, by the families of conditional survivor functions and jump probabilities such that

$$P(\tau_{n+1} > t | F_{\tau_n}) = G_{n, \xi_n}(t) \qquad \text{on } (\tau_n < \infty),$$

$$P(Y_{n+1} = y | F_{\tau_{n+1}^-}) = \pi_{n, \xi_n}(\tau_{n+1}^-, y) \text{ on } (\tau_{n+1} < \infty),$$

where $\xi_n = (\tau_1, \cdots, \tau_n; Y_1, \cdots, Y_n)$, cf. Theorem 1.7.

So far we have mostly discussed CCPE's of class H^E where each of the G is assumed to have a smooth density and the π are assumed to be left-continuous with right-limits as functions of time.

We shall now discuss CCPE's where the G are assumed to be discrete. These processes will prove useful for interpreting estimators in the statistical models we shall be dealing with.

Let Pr be a probability on $(0, \infty]$ and let G be the survivor function for Pr. We shall say that Pr or G is <u>purely discrete</u> if there is a finite or infinite sequence $0 < a_1 < a_2 < \ldots$ of strictly positive and finite numbers such that Pr is concentrated on $\{a_k: k \geq 1\} \cup \{\infty\}$ with $\Pr(\{a_k\}) > 0$ for every k, and such that, if the sequence (a_k) is infinite, $\lim a_k = t^+$, the termination point of Pr. (Since we can order the atoms, Pr cannot be an arbitrary discrete probability).

5.1. <u>Definition</u>. A CCPE, P, belongs to <u>class</u> $\mathcal{D}^E$ if for every $n \geq 0$, $t_1 < \cdots < t_n$, $y_1, \ldots, y_n \in E$, the survivor function $G_{n, t_1 \cdots t_n y_1 \cdots y_n}$ is purely discrete. ▌

Before defining the intensity process for a P of class $\mathcal{D}^E$, we need to discuss intensities for purely discrete probabilities.

If Pr is purely discrete, the <u>intensity</u> function for Pr is $\mu: (0, \infty) \to [0, \infty)$ given by

$$\mu(t) = \begin{cases} g(t)/G(t-) & \text{if } G(t-) > 0 \\ 0 & \text{if } G(t-) = 0, \end{cases}$$

where $g(t) = G(t-) - G(t)$ is the probability $\Pr(\{t\})$. Equivalently if $G(t-) > 0$,

$$\mu(t) = \Pr(\{t\})/\Pr([t\,\rho]).$$

Thus $\mu(t) \leq 1$ always and $\mu(t) > 0$ if and only if t is an atom. The intensity determines G: given μ we get the atoms as the sequence of points $0 < a_1 < a_2 < \dots$ for which $\mu(a_k) > 0$, and then for $k \geq 1$ we have, writing $a_0 = 0$,

$$\mu(a_k) = g(a_k)/G(a_k-)$$

$$= (G(a_{k-1}) - G(a_k))/G(a_{k-1})$$

$$= 1 - \frac{G(a_k)}{G(a_{k-1})}$$

so that

(5.2)
$$G(a_k) = \prod_{j=1}^{k} (1-\mu(a_j)),$$

(5.3)
$$g(a_k) = \mu(a_k) \prod_{j=1}^{k-1} (1-\mu(a_j)).$$

Without referring to the actual values of the a_k, (5.2) and (5.3) can be written in product integral form: introducing the accumulated intensity

$$\mu^{acc}(t) = \sum_{k:\, a_k \leq t} \mu(a_k) = \sum_{s \leq t} \mu(s),$$

it is seen that for $t \geq 0$,

(5.4)
$$G(t) = \prod_{0 < s \leq t} (1-\Delta\mu^{acc}(s))$$

(5.5)
$$g(t) = \Delta\mu^{acc}(t) \prod_{0 < s < t} (1-\Delta\mu^{acc}(s)),$$

(5.6)
$$\frac{g(t)}{G(t-)} = \Delta\mu^{acc}(t),$$

where for any right-continuous function f with limits from the left,
we write $\Delta f(t) = f(t) - f(t-)$ for the size of the discontinuity of f
at t.

The intensities for purely discrete probabilities have one further
property, which together with those given above characterizes the class
of such intensities: by the definition of purely discrete probabilities,
if the sequence (a_k) of atoms is infinite and if $t^{\dagger} < \infty$, then
$\lim a_k = t^{\dagger}$ and $G(t^{\dagger}-) = 0$; therefore by (5.2), $\Sigma \mu(a_k) = \infty$.

Now let $P \in \mathcal{D}^E$ and write $\mu_{nt_1 \ldots t_n y_1 \ldots y_n}$ for the intensity
of $G_{nt_1 \ldots t_n y_1 \ldots y_n}$.

5.7. <u>Definition</u>. The <u>intensity process</u> for $P \in \mathcal{D}^E$ is the stochastic
process $\lambda = (\lambda^y)_{y \in E}$ taking values in $[0,1]^E$ given by

$$\lambda_t^y = \mu_{\overline{N}_{t-} \xi_{\overline{N}_{t-}}}(t) \pi_{\overline{N}_{t-} \xi_{\overline{N}_{t-}}}(t-,y) \qquad (t \geq 0, \ y \in E).$$

The <u>accumulated intensity process</u> for P is the stochastic process
$\Lambda = (\Lambda^y)_{y \in E}$ taking values in $[0,\infty]^E$ given by

$$\Lambda_t^y = \sum_{s \leq t} \lambda_s^y \qquad (t \geq 0, \ y \in E). \qquad\qquad |$$

We shall write $\overline{\lambda} = \Sigma_y \lambda^y$, $\overline{\Lambda} = \Sigma_y \Lambda^y$ in complete analogy with the
case of H^E-processes.

The definition of λ^y should be compared with (2.8). Whereas intensities for processes of class H^E are left-continuous, those for
processes of class $\mathcal{D}^E$ are merely predictable, a concept we shall introduce now and which will prove important for the properties of stochastic integrals.

For $t > 0$, consider the equivalence relation $\underset{t-}{\sim}$ on W^E given
by $w \underset{t-}{\sim} w'$ iff $w(s) = w'(s)$ for all $0 \leq s < t$, and let F_{t-} be
the σ-algebra consisting of all F-measurable unions of equivalence
classes for $\underset{t-}{\sim}$. Of course $F_{t-} \subset F_t$, the inclusion being strict,

since for every $t > 0$ the F_{t-}-atom containing a given path w will include paths from different F_t-atoms, viz. atoms containing paths that jump at time t itself, and atoms of paths that do not. Also define $F_{0-} = F_0 = \{\emptyset, w^E\}$.

Clearly $F_s \subset F_{t-}$ for $s < t$. It can be shown that F_{t-} is the smallest σ-algebra containing all F_s for $s < t$.

Having introduced F_{t-}, we shall call a process $Z = (Z_t)_{t \geq 0}$ predictable if Z_t is F_{t-}-measurable for every $t \geq 0$. (The definition in its final form, appears in Chapter 3, but this one will do for the moment).

We can now state the analogue of Proposition 2.2 for processes of class $\mathcal{D}^E$.

5.8. **Proposition.** The intensity process λ of a CCPE P of class $\mathcal{D}^E$ is predictable and satisfies $\lambda_t^y \geq 0$, $\Sigma_y \lambda_t^y \leq 1$ for all t, y. Also, for P-almost every $w \in W^E$ there is a finite or infinite sequence $t_1 < t_2 < \dots$ of timepoints with $\lim t_k = \infty$ if the sequence is infinite, such that $\overline{\lambda}_t(w) > 0$ if and only if $t = t_k$ for some k. The accumulated intensity Λ has sample paths which are piecewise constant, right-continuous and satisfies $P(\overline{\Lambda}_t < \infty) = 1$ for all t. Moreover Λ is predictable. Finally, for every $t > 0$, $y \in E$

$$(5.9) \qquad P(\Delta N_t^y = 1 \mid F_{t-}) = \lambda_t^y, \qquad P\text{-a.s.}$$

$$(5.10) \qquad P(\Delta \overline{N}_t = 1 \mid F_{t-}) = \overline{\lambda}_t \qquad P\text{-a.s.}$$

Proof. Conditionally on F_{τ_n} inside $(\tau_n < \infty)$, the value of τ_{n+1} is P-almost surely either ∞ or one of the atoms for G_{n, ξ_n}. The fact that G_{n, ξ_n} is purely discrete and Definition 5.7 therefore implies that P-a.s. on $(\tau_{n+1} < \infty)$, $\overline{\lambda}_t > 0$ for only finitely many t in the interval $(\tau_n, \tau_{n+1}]$. Since $\tau_n \uparrow \infty$, this proves the assertion about the discrete structure of λ. That $P(\overline{\Lambda}_t < \infty) = 1$ will follow from

5.12 below. The remaining assertions, apart from (5.9),(5.10) are trivial.

Since only one component of N can jump at the time, (5.10) follows from (5.9) by summation on y. To prove (5.9) observe that on $(\bar{N}_{t-} = n)$, conditioning on F_{t-} is the same as conditioning on F_{τ_n} and the event $(\tau_{n+1} \geq t)$. Therefore on $(\bar{N}_{t-} = n)$

$$
\begin{aligned}
P(\Delta N_t^y = 1 | F_{t-}) &= P(\Delta N_t^y = 1, \tau_{n+1} \geq t | F_{\tau_n}) / P(\tau_{n+1} \geq t | F_{\tau_n}) \\
&= P(\tau_{n+1} = t, Y_{n+1} = y | F_{\tau_n}) / P(\tau_{n+1} \geq t | F_{\tau_n}) \\
&= g_{n, \xi_n}(t) \pi_{n, \xi_n}(t-, y) / G_{n, \xi_n}(t-) \\
&= \lambda_t^y
\end{aligned}
$$

as is seen from Definition 5.7 and (5.6). ∎

The accumulated intensity Λ determines P. To find $G_{nt_1 \ldots t_n y_1 \ldots y_n}(t)$, $\pi_{nt_1 \ldots t_n y_1 \ldots y_n}(t-, y)$ for particular values of n, $t_1, \ldots, t_n, y_1, \ldots, y_n$, t, y, just take a path w such that $N_{t-}(w) = n$, $\tau_k(w) = t_k$, $Y_k(w) = y_k$, $1 \leq k \leq n$, and conclude from Definition 5.7 and (5.4)-(5.6) that

$$
\begin{aligned}
G_{nt_1 \ldots t_n y_1 \ldots y_n}(t) &= \prod_{t_n < s \leq t} (1 - \Delta \bar{\Lambda}_s(w)), \\
\pi_{nt_1 \ldots t_n y_1 \ldots y_n}(t-, y) &= \frac{\Delta \Lambda_t^y(w)}{\Delta \bar{\Lambda}_t(w)} .
\end{aligned}
$$

The following useful result is the analogue of Proposition 2.10 (c).

5.11. __Proposition__. Suppose P is of class $\mathcal{D}^E$ with intensity process λ. Then $\lambda^y(\tau_n^y) > 0$ P-a.s. on $(\tau_n^y < \infty)$ for every $n \in \mathbb{N}_0$, $y \in E$.

__Proof__. We must show that $P(\lambda^y(\tau_n^y) = 0, \ \tau_n^y < \infty) = 0$. But τ_n^y equals one of the jump times τ_k, so it is enough to show that

$$
P(\lambda^y(\tau_n^y) = 0, \ \tau_n^y < \infty, \ \tau_{k+1} = \tau_n^y) = 0
$$

for $k \geq 0$. Conditioning on F_{τ_k} we get

$$P(\lambda^Y(\tau_n^Y) = 0, \ \tau_n^Y < \infty, \ \tau_{k+1} = \tau_n^Y | F_{\tau_k})$$

$$\leq P(\lambda^Y_{\tau_{k+1}} = 0, \ Y_{k+1} = y, \ \tau_{k+1} < \infty | F_{\tau_k})$$

$$= \sum_{s > \tau_k} g_{k,\xi_k}(s) \pi_{k,\xi_k}(s-,y) 1_{(\mu_{k,\xi_k}(s) \pi_{k,\xi_k}(s-,y) = 0)}(s) = 0$$

using (5.5). ▌

The next result is peculiar to the discrete setup.

5.12. <u>Proposition</u>. Suppose P is of class $\mathcal{P}^E$. Then there exists a countable subset D of $(0,\infty]$ such that $P(\tau_n \in D, \ n \geq 1) = 1$. Furthermore, for any $t > 0$ the restriction of P to F_t is discrete, i.e. P is concentrated on a countable collection of F_t-atoms, and if for $n \geq 0$, $t_1 < \ldots < t_n \leq t$, $y_1, \ldots, y_n \in E$, $F = (\bar{N}_t = n, \ \tau_k = t_k, \ Y_k = y_k, \ 1 \leq k \leq n)$ is an arbitrary such F_t-atom, then

$$(5.13) \qquad P(F) = (\prod_{\substack{0 < s \leq t \\ s \neq t_1, \ldots, t_n}} (1 - \Delta \bar{\Lambda}_s (\text{on } F))) \prod_{k=1}^{n} \lambda_{t_k}^{y_k}(\text{on } F).$$

<u>Proof</u>. The possible values for τ_1 under P are the atoms for G_0. Given that τ_1 equals one of these, t, say, only the atoms for $G_{1,t,y}$ where y ranges over E, can occur as values for τ_2. Continuing this argument it is seen that for any n, there is a countable subset D_n of $(0,\infty]$ such that $P(\tau_n \in D_n) = 1$. But then $D = \cup D_n$ is also countable and of course $P(\tau_n \in D, \ n \geq 1) = 1$.

For the second assertion, compute directly

$$P(F) = (\prod_{k=1}^{n} g_{k-1,z_{k-1}}(t_k) \pi_{k-1,z_{k-1}}(t_k-,y_k)) G_{n,z_n}(t)$$

where $z_{k-1} = (t_1, \ldots, t_{k-1}, y_1, \ldots, y_{k-1})$. But by Definition 5.7 and (5.4)-(5.6)

$$(g_{k-1,z_{k-1}}(t_k) \pi_{k-1,z_{k-1}}(t_k-,y_k))/G_{k-1,z_{k-1}}(t_k-) = \lambda_{t_k}^{y_k}(\text{on } F),$$

2.5.7

$$G_{k-1,z_{k-1}}(t_k-) = \underset{t_{k-1}<s<t_k}{\Pi} (1-\Delta\overline{\Lambda}_s(\text{on } F)),$$

$$G_{n,z_n}(t) = \underset{t_n<s\leq t}{\Pi} (1-\Delta\overline{\Lambda}_s(\text{on } F))$$

and (5.13) follows. Finally, since N has only finitely many jumps on $[0,t]$, the possible range of $(\tau_1,\ldots,\tau_{N_t}, Y_1,\ldots,Y_{N_t})$ is countable, from which the discreteness of P restricted to F_t follows. ∎

Remark. We can use the proposition to show that $P(\overline{\Lambda}_t < \infty) = 1$ as asserted in Proposition 5.8. Consider the countably many F_t-atoms F for which $P(F) > 0$. For these the infinite product in (5.13) converges, forcing

$$\underset{\substack{0<s\leq t \\ s \neq t_1,\ldots,t_n}}{\Sigma} \Delta\overline{\Lambda}_s(\text{on } F) < \infty.$$

But the terms omitted for $s = t_1,\ldots,t_n$ are finite (each of them ≤ 1), hence

$$\overline{\Lambda}_t(\text{on } F) = \underset{0<s\leq t}{\Sigma} \Delta\overline{\Lambda}_s(\text{on } F) < \infty.$$

Since the union of the F considered has P-probability one, the desired conclusion emerges. ∎

Remark. Viewing the right hand side of (5.13) as a function of F, hence a function on W^E, we get the likelihood function for observation of N on the time interval $[0,t]$,

$$(5.14) \qquad \ell_t = (\underset{\substack{0<s\leq t \\ s \neq \tau_1,\ldots,\tau_{\overline{N}_t}}}{\Pi} (1-\Delta\overline{\Lambda}_s)) \underset{k=1}{\overset{\overline{N}_t}{\Pi}} \lambda_{\tau_k}^{Y_k} .$$

It is seen that the process $\ell = (\ell_t)_{t\geq 0}$ is predictable with right-continuous, left-limit paths. ∎

We shall now state the analogue of Theorem 2.13 (a) for processes of class $\mathcal{D}^E$. A warning should be issued: statements (b) and (c) of that theorem do not carry over to discrete processes.

5.15. <u>Proposition</u>. Suppose P is a CCPE of class $\mathcal{D}^E$ such that $P\overline{N}_t < \infty$ for all t. Then $P\overline{\Lambda}_t < \infty$ for all t, in fact $P\overline{\Lambda}_t = P\overline{N}_t$, and for every $y \in E$, $M^y = N^y - \Lambda^y$ is a P-martingale.

<u>Proof</u>. By the technique established for processes of class H^E, we need only show that

$$PN^y_{t \wedge \tau_1} = P\Lambda^y_{t \wedge \tau_1}$$

for $t \geq 0$, $y \in E$. But

$$PN^y_{t \wedge \tau_1} = P(\tau_1 \leq t, Y_1 = y)$$

$$= \sum_{s \leq t} g_0(s) \pi_0(s-,y)$$

and

$$P\Lambda^y_{t \wedge \tau_1} = P \sum_{s \leq t \wedge \tau_1} \mu_0(s) \pi_0(s-,y)$$

$$= (\sum_{s \leq t} \mu_0(s) \pi_0(s-,y)) G_0(t) + \sum_{u \leq t} g_0(u) \sum_{s \leq u} \mu_0(s) \pi_0(s-,y).$$

Interchanging the order of summation in the last term reduces that to

$$\sum_{s \leq t} \mu_0(s) \pi_0(s-,y) \sum_{s \leq u \leq t} g_0(u)$$

$$= \sum_{s \leq t} \mu_0(s) \pi_0(s-,y) (G_0(s-) - G_0(t)),$$

so that by (5.6)

$$P\Lambda^y_{t \wedge \tau_1} = \sum_{s \leq t} \mu_0(s) \pi_0(s-,y) G_0(s-)$$

$$= \sum_{s \leq t} g_0(s) \pi_0(s-,y)$$

as required. ∎

The intensity λ of a $P \in \mathcal{P}^E$ satisfies $\Delta \overline{\lambda}_t = \overline{\lambda}_t \leq 1$ as stated in Proposition 5.8. To understand better some estimation results in the statistical theory, we shall however need processes with intensities that do not satisfy this constraint. The natural way to accomplish this is to introduce (discrete) counting processes with multiple jumps. In some sense such a process can be thought of as the limit of a sequence of CCPE's, arranged so that some of the jumps of a member far out in the sequence occur closely together, and in the limit collapse to one "big" jump.

Therefore, consider the space $\widetilde{W}^E$ of paths $w: [0,\infty) \to \mathbb{N}_0^E$ which are right-continuous, each component increasing from 0 in jumps of size $1,2,\ldots,$ and where more components may jump simultaneously.

The path-space $\widetilde{W}^E$ can be equipped with a filtered measurable structure in an obvious manner, and we can then define a multivariate multijump canonical counting process with type-set E as a probability $\widetilde{P}$ on $\widetilde{W}^E$.

But in a straightforward manner such a process may be viewed as an (ordinary) multivariate canonical counting process with countably infinite type-set

$$\widetilde{E} = \{\widetilde{x} = (x^y)_{y \in E} \in \mathbb{N}_0^E : \sum_y x^y \geq 1\},$$

viz., let $\widetilde{N}_t: \widetilde{W}^E \to \mathbb{N}_0^E$ be the projection $\widetilde{N}_t(\widetilde{w}) = \widetilde{w}_t,$ and then for $\widetilde{x} \in \widetilde{E}$ define $K = (K^{\widetilde{x}})_{\widetilde{x} \in \widetilde{E}}$ by

$$K_t^{\widetilde{x}} = \sum_{s \leq t} 1_{(\Delta \widetilde{N}_s^y = x^y, \; y \in E)},$$

i.e. $K_t^{\widetilde{x}}$ is the number of times in $(0,t]$ that, simultaneously for all $y \in E$, the y'th component in $\widetilde{N}$ has suffered a jump of size x^y. Obviously K is a counting process in the sense of Definition 1.1 (except that the type-set is countably infinite, but we shall not worry about this). Clearly the behaviour of K on $[0,t]$ is determined by the behaviour of $\tilde{N}$ on $[0,t]$. Conversely knowledge of K on $[0,t]$ determines $\tilde{N}$ on $[0,t]$ since

(5.16)
$$\tilde{N}_t^y = \sum_{\tilde{x}\in\tilde{E}} x^y K_t^{\tilde{x}} .$$

Therefore, instead of studying $\tilde{P}$, we may as well study the CCP$\tilde{E}$ generated by K.

We shall only need discrete counting processes with multiple jumps. Therefore, let now P be a CCP$\tilde{E}$ of class $\mathcal{D}^{\tilde{E}}$, and let $\lambda = (\lambda^{\tilde{x}})_{\tilde{x}\in\tilde{E}}$ and $\Lambda = (\Lambda^{\tilde{x}})_{\tilde{x}\in\tilde{E}}$ denote the intensity and the accumulated intensity, i.e.

$$\lambda_t^{\tilde{x}} = P(\Delta\tilde{N}_t^y = x^y,\ y\in E|F_{t-}), \qquad \Lambda_t^{\tilde{x}} = \sum_{s\le t}\lambda_s^{\tilde{x}} ,$$

where, to have everything defined on W^E and in accordance with (5.16)

$$\tilde{N}_t^y = \sum_{\tilde{x}\in\tilde{E}} x^y N_t^{\tilde{x}} \qquad (t\ge 0,\ y\in E).$$

We define the intensity $\tilde{\lambda} = (\tilde{\lambda}^y)_{y\in E}$ for $\tilde{N}$ by

(5.17)
$$\tilde{\lambda}_t^y = \sum_{\tilde{x}\in\tilde{E}} x^y \lambda_t^{\tilde{x}} \qquad (t\ge 0,\ y\in E),$$

and it is quite clear that although $\sum_{\tilde{x}}\lambda_t^{\tilde{x}}\le 1$, there is no limitation on the size of the $\tilde{\lambda}^y$, in fact without conditions on P, they may be infinite.

Using (5.9) it is seen that

$$\tilde{\lambda}_t^y = \sum_{\tilde{x}} x^y P(\Delta N_t^{\tilde{x}} = 1|F_{t-})$$

$$= \sum_{\tilde{x}} x^y P(\Delta\tilde{N}_t^z = x^z,\ z\in E|F_{t-}) .$$

Here we need only consider $\tilde{x}$ such that $x^y\ge 1$, but then (cf. the definition of $\tilde{E}$) the x^z for $z\ne y$ vary freely over $\mathbb{N}_0$ and therefore

$$\tilde{\lambda}_t^y = \sum_{x^y} x^y P(\Delta\tilde{N}_t^y = x^y|F_{t-})$$

$$= P(\Delta\tilde{N}_t^y|F_{t-}) .$$

It is also seen from Proposition 5.15 that if $P\tilde{N}_t^y < \infty$ for all t, then $\tilde{N}^y - \tilde{\Lambda}^y$ is a P-martingale, where of course $\tilde{\Lambda}_t^y = \sum\limits_{s \leq t} \tilde{\lambda}_s^y$.

It is important to observe that the intensity $\tilde{\lambda}$ or the accumulated intensity $\tilde{\Lambda}$ does not determine the P-distribution of $\tilde{N}$ (equivalently, does not determine P itself), contrary to the case of ordinary CCPE's. Formally this amounts to saying that the $\lambda^{\tilde{x}}$ cannot be found from the $\tilde{\lambda}^y$ using (5.17).

In the one-dimensional case, i.e. if E consists of just one element, the preceding simplifies a great deal. Then $\tilde{E} = \mathbb{N}$ and for $\tilde{x} \in \tilde{E}$, $N_t^{\tilde{x}}$ counts the number of jumps of size $\tilde{x}$ for the original multiple jump process. The intensity that identifies the CCP$\tilde{E}$ is $(\lambda^{\tilde{x}})_{\tilde{x} \in \tilde{E}}$ where

$$\lambda_t^{\tilde{x}} = P(\Delta \tilde{N}_t = x \mid F_{t-}) \ ,$$

while the intensity $\tilde{\lambda}$ is

$$\tilde{\lambda}_t = \sum\limits_{\tilde{x} \in \tilde{E}} \tilde{x} \lambda_t^{\tilde{x}} \ ,$$

which does not determine the CCP$\tilde{E}$.

An instance of such a one-dimensional multiple jump process was given in Example 1.21.

Exercises.

1. Let P be a CCPE of class H^E such that all $G_{nt_1 \cdots t_n y_1 \cdots y_n}$ have termination point ∞. Show, using the description of inhomogeneous Markov chains from Example 2.1.19, that with respect to P, the process N is a Markov chain (with state-space $\mathbb{N}_0^E$) if and only if for every t, P-a.s the intensity λ_{t-} is a function of N_{t-} alone. ▐

2. Under the assumptions of Theorem 2.2.13, and with the notation from there, show that

$$P \, M_{t \wedge \tau_1}^{y2} = P \, \Lambda_{t \wedge \tau_1}^{y} \, . \qquad\qquad ▐$$

3. Show by computation that it is not true that the identity in Exercise 2 holds for all P of class $\mathcal{D}^E$. ▐

3. STOCHASTIC INTEGRALS

3.1. <u>Processes and martingales on</u> W^E.

Before coming to the discussion of stochastic integrals, it will be useful to introduce some terminology from the general theory of stochastic processes. Some of the concepts have been defined earlier in these notes. We shall repeat the definitions for the sake of completeness.

In the general theory one is given a filtered space $(\Omega, A, A_t, \mathbb{P})$ satisfying what is called "the usual conditions", i.e. apart from the condition that $A_s \subset A_t$ for $s \leq t$, the sub σ-algebras A_t of A satisfies that $A_{t+} \overset{D}{=} \underset{\varepsilon>0}{\cap} A_{t+\varepsilon}$ equals A_t' for all t and also every A_t is complete so that if $N \in A$ with $\mathbb{P}(N) = 0$, then $N' \in A_t$ for all t where N' is an arbitrary subset of N.

We shall base our discussion on the filtered space (W^E, F, F_t, P) where P is a given CCPE on the path-space W^E of counting process paths with type set E. Some definitions below only involve W^E and the measurable structure on W^E, while others depend on P as well.

It must be emphasized that the canonical filtration on W^E does not satisfy the "usual conditions": the definition of F, F_t has nothing to do with the probability P, and in particular the σ-algebras have not been completed to include subsets of P-null sets. This means that some of the definitions below, although modeled on the corresponding definitions in the general theory, will differ slightly.

Let now W^E and P be given and consider a real-valued process $Z = (Z_t)_{t \geq 0}$ defined on (W^E, F). Recall that Z is <u>adapted</u> if Z_t is F_t-measurable for every t.

1.1. <u>Definition</u>. A real-valued process Z is <u>measurable</u> if the mapping $(t, w) \to Z_t(w)$ from $[0, \infty) \times W^E$ to $\mathbb{R}$ is measurable with respect to the product σ-algebra $B \otimes F$ on $[0, \infty) \times W^E$, where B is the Borel subsets of $[0,\infty)$.   ∎

A simple conditon for a process to be measurable is the following:

1.2. <u>Proposition</u>. A real-valued process Z with sample paths that
are either everywhere right-continuous or everywhere left-continuous,
is measurable.

<u>Proof</u>. We just consider the right-continuous case. Then for every
$t \geq 0$

$$Z_t(w) = \lim_{n \to \infty} \sum_{k=1}^{\infty} Z_{\frac{k}{2^n}}(w) \; 1_{[\frac{k-1}{2^n}, \frac{k}{2^n})}(t)$$

represents $Z_t(w)$ as a limit of functions on $[0,\infty) \times W^E$ which are
obviously measurable.   ∎

1.3. <u>Definition</u>. An <u>increasing process</u> is an adapted $[0,\infty]$-valued
stochastic process $A = (A_t)$ such that $A_0 = 0$ and such that the
mapping $t \to A_t(w)$ is right-continuous and non-decreasing for all
$w \in W^E$.

An increasing process A is <u>P-finite</u> if $P(A_t < \infty) = 1$ for all
$t \geq 0$, <u>P-locally integrable</u> if $PA_t < \infty$ for all $t \geq 0$ and <u>P-inte-
grable</u> if $PA_\infty < \infty$ where $A_\infty \overset{D}{=} \lim_{t\uparrow\uparrow\infty} A_t$.   ∎

Obviously "P-integrable" implies "P-locally integrable" implies
"P-finite". Since by definition a CCPE P is stable, the increasing
process N^y is P-finite for all y. If P is of class H^E, the
integrated intensity Λ^y is increasing, and by theorem 2.2.13, if
N^y is P-locally integrable, so is Λ^y. (To say, as was done in Section 2.2, that P has finite expectations locally or that each N^y is
P-locally integrable, is of course the same thing. But having "finite
expectations locally" is a property of a probability, while being

"locally integrable" is a property of a stochastic process).

In Section 2.5 we introduced for $t > 0$, F_{t-} as the σ-algebra comprising F-measurable unions of equivalence classes for the equivalence relation $\underset{t-}{\sim}$ on W^E given by $w \underset{t-}{\sim} w'$ iff $w(s) = w'(s)$ for all $0 \leq s < t$. We also defined $F_{0-} = \{\emptyset, W^E\}$.

1.4. Definition. A real-valued process Z is __predictable__ if it is measurable and Z_t is F_{t-}-measurable for every $t \geq 0$. ▮

In particular a predictable process is adapted.

1.5. Proposition. A real-valued process Z which is adapted and has left-continuous sample paths, is predictable.

__Proof.__ By Proposition 1.2, Z is measurable, and since $F_{0-} = F_0$, Z_0 is F_{0-}-measurable because Z is adapted. For $t > 0$, let $t_n < t$ be a sequence increasing to t. By left-continuity $Z_t = \lim Z_{t_n}$, and since for every n, Z_{t_n} is F_{t_n} - and hence F_{t-}-measurable, Z_t is F_{t-}-measurable. ▮

Notice that the processes N^y are not predictable, while the integrated intensities Λ^y for counting processes of class H^E, are predictable because they have continuous paths. Nice examples of predictable processes with right-continuous but not left-continuous paths, are provided by the accumulated intensities for counting processes of class $\mathcal{D}^E$, see Section 2.5.

1.6. Definition. A real-valued measurable process Z is __P-evanescent__ if there exists $N \in F$ with $P(N) = 0$ such that $\underset{t>0}{U} (Z_t \neq 0) \subset N$. Two measurable processes Z, Z' are __P-indistinguishable__ if the difference $Z-Z'$ is P-evanescent. ▮

Remark. Notice that if Z, Z' both have right-continuous (or left-continuous) sample paths, they are indistinguishable if $P(Z_t = Z_t') = 1$ for all $t \geq 0$ because

$$\bigcup_{t \geq 0} (Z_t = Z_t') = \bigcup_{q \geq 0} (Z_q = Z_q') \quad ,$$

where q ranges over the non-negative rationals. ∎

1.7. Definition. A real-valued process $M = (M_t)$ is a P-martingale if M is adapted with all sample paths right-continuous, $P|M_t| < \infty$ for all $t \geq 0$ and $P(M_t | F_s) = M_s$ for all $0 \leq s \leq t$. The process M is a P-submartingale if instead of the last equality only the inequality $P(M_t | F_s) \geq M_s$ holds. ∎

Any increasing process which is adapted and P-locally integrable, is automatically a submartingale.

By the martingale convergence theorem, any submartingale (which is right-continuous by definition) has sample paths which have left-limits everywhere, the limits being limits in $\mathbb{R}$. If M is a P-martingale such that either $\sup_{t>0} PM_t^+ < \infty$ or $\sup_{t>0} PM_t^- < \infty$, or if M is a P-submartingale such that $\sup_{t>0} PM_t^+ < \infty$, then $M_\infty = \lim_{t \uparrow \uparrow \infty} M_t$ exists P-a.s. and $P(|M_\infty| < \infty) = 1$. (Here $M_t^+ = M_t \vee 0$, $M_t^- = -(M_t \wedge 0)$).

A P-submartingale M is uniformly integrable if $\lim_{a \uparrow \uparrow \infty} \sup_{t \geq 0} P(|M_t| ; |M_t| \geq a) = 0$. By the remarks above M_∞ then exists and is almost surely finite. But even more is true: $P|M_\infty| < \infty$, $P(M_\infty | F_t) = M_t$ for all t if M is a martingale, $P(M_\infty | F_t) \geq M_t$ if M is a submartingale, and $\lim_{t \uparrow \uparrow \infty} P|M_\infty - M_t| = 0$.

Let M be a P-martingale such that M_∞ exists and is finite almost surely, and let $\sigma \leq \tau$ be two stopping times. Since M_∞ exists, M_σ and M_τ are well defined on all of W^E (also on the sets $(\sigma = \infty)$, $(\tau = \infty)$). We shall say that the optional sampling theorem holds for σ and τ if $P(M_\tau|F_\sigma) = M_\sigma$.

The optional sampling theorem always holds if τ is bounded (i.e. $\sup \tau(w) < \infty$ and in that case M_∞ need not exist), or for any stopping times $\sigma \leq \tau$ provided M is uniformly integrable.

1.8. <u>Definition</u>. A real-valued process M is a <u>local P-martingale</u> if it is adapted and there exists a sequence $(\sigma_n)_{n\geq 1}$ of stopping times with $\sigma_n(w) \uparrow \infty$ for P-almost all w such that for every n, the process $(M_{t \wedge \sigma_n})_{t \geq 0}$ is a P-martingale.   ∎

We have introduced local martingales merely to be able to quote correctly some results from the general theory, but shall not need the definition otherwise. (In our setup, the sequence (τ_n) of jump times is an obvious candidate for the (σ_n) in the definition).

Any martingale is a local martingale: for every $n \geq 1$, $t \geq 0$, $t \wedge \sigma_n$ is a bounded stopping time, so by optional sampling, $P(M_{t \wedge \sigma_n}|F_{s \wedge \sigma_n}) = M_{s \wedge \sigma_n}$ for $s \leq t$. But on $(\sigma_n > s)$ conditioning on $F_{s \wedge \sigma_n}$ is the same as conditioning on F_s , so the identity becomes $P(M_{t \wedge \sigma_n}|F_s) = M_{s \wedge \sigma_n}$, and this holds trivially on $(\sigma_n \leq s)$. Thus $(M_{t \wedge \sigma_n})$ is a martingale, no matter how σ_n is chosen.

We shall denote by $M(P)$ the space of all P-martingales. Furthermore we shall write $M_0^2(P)$ for the space of <u>locally square integrable</u> P-martingales, i.e. $M \in M_0^2(P)$ if M is a P-martingale and $PM_t^2 < \infty$ for all t .

With this notation, Theorem 2.2.13 may be stated as follows: if P is a CCPE such that each N^y is P-locally integrable, then also Λ^y is P-locally integrable and $M^y = N^y - \Lambda^y \in M_0^2(P)$, $M^{y2} - \Lambda^y \in M(P)$ for all y while $M^y M^z \in M(P)$ for all $y \neq z$.

We shall now discuss the <u>Doob-Meyer decomposition</u> theorem for submartingales.

1.9. <u>Theorem</u>. Suppose Z is a P-submartingale. Then there exists a
local martingale M and a P-locally integrable, predictable, increasing process A such that

(1.10) $$Z = M + A .$$

Moreover, if Z = M' + A' is another such decomposition, then the
processes M,M'(and A,A') are indistinguishable. |

The identity (1.10) states that $Z_t(w) = M_t(w) + A_t(w)$ for all
t,w and not just that Z = M + A P-a.s. It is easy to see that from
a decomposition Z = M + A valid P-a.s. one can get another valid
everywhere: simply define $M^* = Z - A$. Then $M^* = M$ P-a.s., M^*
is adapted and locally P-integrable (because Z and A are), and consequently M^* is also a P-martingale.

We shall not establish the existence of the decomposition in complete generality, but only in the cases of interest to us. (Examples
of decompositions have already been given in Theorems 1.5.1, 2.2.13
and Proposition 2.5.15). However, we shall prove the uniqueness below.

The theorem applies of course in particular if Z is a locally
P-integrable, adapted, increasing process. In that case the A from
the decomposition is often referred to as the <u>compensator</u> for Z . Also
A is then what is called the <u>dual predictable projection</u> of Z .

<u>Remark</u>. It can be shown that the predictable increasing process A
from the decomposition of Z has an additional property, namely it is
<u>natural</u>. This means that

(1.11) $$P\int_0^t dA_s \tilde{M}_s = P\int_0^t dA_s \tilde{M}_{s-}$$

for every t $\geq$ 0 and every bounded P-martingale $\tilde{M}$. We shall not need
this property. The integrals appearing in (1.11) of $\tilde{M}$ with respect
to A are, for w fixed, ordinary Lebesgue-Stieltjes integrals of the
type we shall discuss in detail in the next   section. Notice that

(1.11) is automatic if A is continuous.   ▌

Predictable processes are important, not only to give uniqueness
of the decomposition in Theorem 1.9, but also in the theory of stochastic integration, as we shall presently see. One fundamental property
is the following.

1.12.$\underline{\text{Proposition}}$. Suppose M is a predictable local P-martingale with
$M_0 = 0$.   Then M is evanescent.

$\underline{\text{Proof}}$. We first claim that it suffices to consider martingales. Suppose the result has been proved for martingales, and let M be a local
P-martingale. Let $\sigma_n \uparrow \infty$ be a sequence of stopping times such that
for every n , $M^{(n)} = (M_{t \wedge \sigma_n})_{t \geq 0}$ is a P-martingale. Obviously
$M_0^{(n)} = 0$,   and $M^{(n)}$ is predictable, because if $t > 0$ and $w \underset{t-}{\sim} w'$,
then since $(\sigma_n < t) = \underset{k \geq 1}{U} (\sigma_n \leq t - \frac{1}{k}) \in F_{t-}$, $\sigma_n(w) < t$ iff $\sigma_n(w') < t$,
and so $M_t^{(n)}(w) = M_t^{(n)}(w')$ follows because M is predictable. As the
proposition is supposed to hold for martingales, it follows therefore
that each $M^{(n)}$ is evanescent. But then we can find $N \in F$ with
$P(N) = 0$ such for $w \notin N$, $M_t^{(n)}(w) = 0$ simultaneously for all n and
t . Since $M_t = \lim_{n \to \infty} M_t^{(n)}$, M is also evanescent.

We next claim that for every $n \geq 0$, $0 < t_1 < \cdots < t_n$, $y_1, \cdots, y_n \in E$,
writing $z_n = (t_1, \cdots, t_n; y_1, \cdots, y_n)$,   there is a right-continuous
function $f_n(z_n; \cdot): (t_n, \infty) \to \mathbb{R}$ such that

(1.13) $M_t = f_n(\xi_n; t)$ on $(\tau_n < t \leq \tau_{n+1})$,

where as usual $\xi_n = (\tau_1, \cdots, \tau_n; Y_1, \cdots, Y_n)$. (For $n = 0$, f_0 is
just a function of t . It is critical that (1.13) holds for $t = \tau_{n+1}$,
and it is for this we shall use the assumption that M is predictable).
To obtain (1.13), simply define

$$f_n(z_n; t) = M_t(w)$$

for any path w such that $\tau_n(w) < t \leq \tau_{n+1}(w)$ and $\xi_n(w) = z_n$. The definition is consistent because if w' is another such path, then $w \underset{\widetilde{t-}}{\sim} w'$ wherefore $M_t(w) = M_t(w')$. It is clear that $f_n(z_n; \cdot)$ is defined on all of (t_n, ∞), and it is right-continuous since M is. As a special case of (1.13) we have

$$(1.14) \qquad M_{\tau_{n+1}} = f_n(\xi_n; \tau_{n+1}) \qquad \text{on } (\tau_{n+1} < \infty).$$

Now, by optional sampling

$$(1.15) \qquad P(M_{\tau_{n+1} \wedge t} | F_{\tau_n}) = P(M_{\tau_{n+1} \wedge t} | F_{\tau_n \wedge t}) = M_{\tau_n \wedge t} = M_{\tau_n}$$

P-a.s. on $(\tau_n < t)$. But if we write μ_n for the intensity function μ_{n, ξ_n} of the conditional distribution of τ_{n+1} given τ_n, the left hand side becomes

$$P(f_n(\xi_n; \tau_{n+1}); \ (\tau_{n+1} \leq t) | F_{\tau_n}) + P(f_n(\xi_n; t); \ (\tau_{n+1} > t) | F_{\tau_n})$$

$$= \int_{\tau_n}^t ds \, \mu_n(s) \exp\left(-\int_{\tau_n}^s \mu_n\right) f_n(\xi_n; s) + \exp\left(-\int_{\tau_n}^t \mu_n\right) f_n(\xi_n; t)$$

and (1.15) may be written

$$(1.16) \quad \int_{\tau_n}^t ds \, \mu_n(s) \exp\left(-\int_{\tau_n}^s \mu_n\right) f_n(\xi_n; s) + \exp\left(-\int_{\tau_n}^t \mu_n\right) f_n(\xi_n; t) = M_{\tau_n} .$$

At first, being an equality between conditional expectations, this identity holds P-a.s. on $(\tau_n < t)$ for every fixed t, but since everything in sight is right-continuous as a function of t, we can get the identity P-a.s. on $(\tau_n < \infty)$, simultaneously for all $t > \tau_n$.

The proposition will now follow if we show the following: if μ is the intensity for a probability on (a, ∞), where $a \geq 0$, with termination point $t^\dagger$, and if $g: (a, \infty) \to \mathbb{R}$ is right-continuous such that

$$(1.17) \qquad \int_a^t ds \, \mu(s) e^{-\int_a^s \mu} g(s) + \left(e^{-\int_a^t \mu}\right) g(t) = c$$

for all $t > a$, where c is a constant, then $g \equiv c$ on $(t, t^\dagger)$. Indeed, applying (1.17) to (1.16), gives, since P-a.s. $\tau_{n+1} < \tau_n^\dagger$ on $(\tau_n^\dagger < \infty)$ with $\tau_n^\dagger$ the termination point for the distribution with intensity μ_n, that P-a.s. M is constant on each of the intervals $(\tau_n, \tau_{n+1}] \setminus \{\infty\}$ as long as $\tau_n < \infty$, and equal to M_{τ_n} on that interval, in particular $M_{\tau_{n+1}} = M_{\tau_n}$ on $(\tau_{n+1} < \infty)$. Since $M_{\tau_0} = M_0 = 0$, M is therefore evanescent.

It remains to prove (1.17). Let $t < t^\dagger$ amd multiply both sides of (1.17) by $\exp(\int_a^t \mu)$ (which is finite), and isolate g to see firstly that g is continuous and secondly that g is differentiable from the right. Differentiating directly in (1.17) gives

$$\left(e^{-\int_a^t \mu} \right) (D^+ g)(t) = 0$$

so that $D^+ g \equiv 0$ on $(a, t^\dagger)$. Inserting a constant for g in (1.17) now shows that $g \equiv c$ on $(a, t^\dagger)$. |

Remark. The proof of the proposition relies very much on the structure of the path-space W^E. As it stands, the proposition is not true in the genreal theory of processes, the one-dimensional Brownian motion being an example of a non-evanescent, continuous (hence predictable) martingale. However, if it is also assumed that the sample paths of the local martingale M are of bounded variation on every finite time interval, the proposition remains true also in the general case. |

Proof of the uniqueness asertion of Theorem 1.9. Suppose $Z = M + A = M' + A'$ are two decompositions of the submartingale Z into a local martingale and a locally integrable, predictable, increasing process. Then $M - M' = A - A'$ is a predictable local martingale starting at 0. By Proposition 1.12 therefore $M - M'$ and $A - A'$ are evanescent. |

Consider a locally square-integrable P-martingale $M \in M_0^2(P)$. Then M^2 is a P-submartingale and by Theorem 1.9 can be represented as the sum of a local P-martingale and a locally integrable, predictable, increasing process. We shall denote the latter by $<M,M>$, or just by $<M>$. Notice that $<M>_0 = 0$.

If $M^{(1)}, M^{(2)} \in M_0^2(P)$ also $M^{(1)} \pm M^{(2)} \in M_0^2(P)$ and we define

$$<M^{(1)}, M^{(2)}> = \frac{1}{4}(<M^{(1)} + M^{(2)}> - <M^{(1)} - M^{(2)}>).$$

Clearly $<M^{(1)}, M^{(2)}>$ is locally P-integrable and predictable and $<M^{(1)}, M^{(2)}> = <M^{(2)}, M^{(1)}>$. We say that $M^{(1)}$ and $M^{(2)}$ are <u>orthogonal</u> if $<M^{(1)}, M^{(2)}>$ is evanescent.

1.18.<u>Proposition</u>. If $M^{(1)} M^{(2)} \in M_0^2(P)$, then $<M^{(1)}, M^{(2)}>$ is up to indistinguishability the unique locally P-integrable and predictable process starting at 0 such that $M^{(1)} M^{(2)} - <M^{(1)}, M^{(2)}>$ is a local P-martingale. In particular two martingales $M^{(1)}, M^{(2)} \in M_0^2(P)$ are orthogonal iff the product $M^{(1)} M^{(2)}$ is a local martingale.

<u>Proof</u>. Only the first assertion requires proof, the second being an easy consequence. For simplicity we shall assume that $M^{(1)} - <M^{(1)}>$ and $M^{(2)} - <M^{(2)}>$ are martingales rather than local martingales, and then show first that $M^{(1)} M^{(2)} - <M^{(1)}, M^{(2)}>$ is a martingale. But for $s \leq t$

$$P(M_t^{(1)} M_t^{(2)} - <M^{(1)}, M^{(2)}>_t | F_s)$$

$$= \frac{1}{4} P([(M_t^{(1)} + M_t^{(2)})^2 - <M^{(1)} + M^{(2)}>_t] - [(M_t^{(1)} - M_t^{(2)})^2 - <M^{(1)} - M^{(2)}>_t] | F_s)$$

$$= \frac{1}{4}([(M_s^{(1)} + M_s^{(2)})^2 - <M^{(1)} + M^{(2)}>_s] - [(M_s^{(1)} - M_s^{(2)})^2 - <M^{(1)} - M^{(2)}>_s])$$

$$= M_s^{(1)} M_s^{(2)} - <M^{(1)}, M^{(2)}>_s$$

as desired.

If B is another locally integrable and predictable process starting at 0 such that $M^{(1)}M^{(2)} - B$ is a martingale, then $B - \langle M^{(1)}, M^{(2)} \rangle$ is a predictable martingale starting at 0 . Hence $B = \langle M^{(1)}, M^{(2)} \rangle$ P-a.s. by Proposition 1.12. $\blacksquare$

1.19. <u>Proposition</u>. If $M^{(1)}, M^{(2)}, M^{(3)} \in M_0^2(P)$, then

$$\langle \alpha M^{(1)} + \beta M^{(2)}, M^{(3)} \rangle = \alpha \langle M^{(1)}, M^{(3)} \rangle + \beta \langle M^{(2)}, M^{(3)} \rangle .$$

<u>Proof</u>. By Proposition 1.18 we need only show that

$$(\alpha M^{(1)} + \beta M^{(2)}) M^{(3)} - [\alpha \langle M^{(1)}, M^{(3)} \rangle + \beta \langle M^{(2)}, M^{(3)} \rangle]$$

is a martingale. Rearranging the terms as

$$\alpha (M^{(1)} M^{(3)} - \langle M^{(1)}, M^{(3)} \rangle) + \beta (M^{(2)} M^{(3)} - \langle M^{(2)}, M^{(3)} \rangle)$$

this is obvious, again using Proposition 1.18. $\blacksquare$

Applying the angle brackets notation to Theorem 2.2.13, we see that (b), (c) states that $M^y \in M_0^2(P)$ for all y with $\langle M^y \rangle = \Lambda^y$, $\langle M^y, M^z \rangle = 0$ if $y \neq z$.

3.2. Definition and basic properties of stochastic integrals.

We shall discuss stochastic integrals of processes defined on the path-space W^E. Such integrals involve two stochastic processes: one process is to be integrated with respect to the other. The resulting stochastic integral becomes a stochastic process itself.

Let A be an increasing process. For every $w \in W^E$ the increasing function $t \to A_t(w)$ induces a measure $\mu(w, \cdot)$ on $[0, \infty)$, which is certainly positive, but may give measure $+\infty$ to some bounded subsets of $[0, \infty)$. If we define $a = \inf\{t > 0: A_t = \infty\}$, then $\mu(w, B) < \infty$ for all Borel subsets of $[0, a(w))$ which are bounded away from $a(w)$, and in particular $\mu(w, (s,t]) = A_t(w) - A_s(w)$ whenever $s \leq t < a(w)$. On the interval $[a(w), \infty)$, $\mu(w, \cdot)$ is given by $\mu(w, (a(w), \infty)) = 0$ while $\mu(w, \cdot)$ has an atom at $a(w)$ iff $\lim_{t \uparrow\uparrow a} A_t(w) < \infty$ in which case $\mu(w, \{a(w)\}) = \infty$. Finally notice that $\mu(w, \{0\}) = 0$ since $\lim_{t \downarrow\downarrow 0} A_t(w) = 0$ by right-continuity.

If $f: [0, \infty) \to \mathbb{R}$ is measurable and non-negative, the integral $\int \mu(w, ds) f(s)$ is well-defined: on the interval $[0, a(w))$ the integral is an ordinary Lebesgue-Stieltjes integral. In the sequel we shall denote the integral by $\int dA_s(w) f(s)$ and write $\int_{(0,t]}$ for the integral from 0 up to and including t. (As noted above it does not matter whether 0 is included in the domain of integration).

In applications $A_t(w)$ will almost always be finite. More precisely, with a probability P given on (W^E, F), we shall assume that A is P-finite. It is however convenient to have $\int_{(0,t]} dA_s(w) f(s)$ defined for all t, A increasing, $f \geq 0$ measurable and all w without reference to a probability.

2.1. Definition. For A an increasing process and $Z \geq 0$ a real-valued measurable process, the stochastic integral of Z with respect to A is defined as the family $A(Z) = (A_t(Z))_{t \geq 0}$ of $[0, \infty]$-valued functions on W^E given by

(2.2) $$A_t(Z) = \int_{(0,t]} dA_s \, Z_s \; . \qquad \blacksquare$$

The assumption that Z be measurable ensures in particular that $s \rightarrow Z_s(w)$ is measurable for all w, so the integral (2.2) is defined everywhere on W^E.

Before listing some basic properties of stochastic integrals, we need one new concept.

2.3. <u>Definition</u>. A real-valued process $Z = (Z_t)_{t \geq 0}$ is <u>locally uniformly bounded</u> if

$$\sup_{s \leq t, w \in W^E} |Z_s(w)| < \infty$$

for every $t \geq 0$. A process Z is <u>uniformly bounded</u> if

$$\sup_{s \geq 0, w \in W^E} |Z_s(w)| < \infty \; . \qquad \blacksquare$$

2.4. <u>Proposition</u>. The stochastic integral $A(Z)$ of a measurable process $Z \geq 0$ with respect to an increasing process A satisfies

(i)   the family $(A_t(Z))_{t \geq 0}$ is a non-negative, right-continuous stochastic process with increasing sample paths, starting at 0, taking values in $[0, \infty]$;

(ii)   the process $A(Z)$ is adapted (predictable) if both A and Z are adapted (predictable);

(iii)   if $Z' \geq 0$ is another measurable process, then
        $A(\alpha Z + \alpha' Z') = \alpha A(Z) + \alpha' A(Z')$ for all $\alpha, \alpha' \geq 0$;

(iv)   if A' is another increasing process, then
        $(\alpha A + \alpha' A')(Z) = \alpha A(Z) + \alpha' A'(Z)$ for all $\alpha, \alpha' \geq 0$;

(v)   if $(Z^{(n)})_{n \geq 1}$ is a sequence of measurable processes $Z^{(n)} \geq 0$, increasing to Z pointwise, i.e. $Z_t^{(n)}(w) \uparrow Z_t(w)$ for all t, w, then $A_t(Z^{(n)})(w) \uparrow A_t(Z)(w)$ for all t, w ;

(vi) if Z is locally uniformly bounded, then $A_t(Z) < \infty$ on $(a > t)$.

Proof. For every w with $t < a(w)$, $A_t(Z)(w)$ is a Lebesgue-Stieltjes integral, and most of the assertions are elementary properties of such integrals. Therefore we shall only prove (i) and (ii).

In (i) the only non-obvious claim is that $A(Z)$ is a process, i.e. each $A_t(Z)$ is F-measurable. To prove this, observe first that $Z_t(w)$ as a measurable function of t and w can be approximated from below by an increasing sequence of finite sums, where each term is a constant times the indicator function 1_C of a measurable subset C of $[0,\infty) \times W^E$. Thus, using (iii) and (v), (i) will be true if the measurability of $A_t(Z)$ is shown for indicator processes 1_C. But the class of sets C for which $A_t(1_C)$ is measurable for all t is closed under the formation of finite disjoint unions and monotone increasing limits. Hence it is enough to consider C of the form $C = I \times F$ where I is an interval and $F \in F$, and in this case the measurability is checked immediately from the definition of the stochastic integral.

To prove (ii), observe that if e.g. A and Z are adapted, then for every t , $A_t(Z)$ is constant on F_t-atoms, and since by (i) it is measurable it is also F_t-measurable. |

The assumption in (vi) that Z be locally uniformly bounded, is far too much to ensure the conclusion. (That $\sup\limits_{s \leq t} Z_s(w) < \infty$ would suffice). However, we shall chiefly work with locally uniformly bounded integrands later, and have therefore chosen the formulation above.

Since $A_t(Z)$ increases with t , we can define $A_\infty(Z) = \lim\limits_{t\uparrow\infty} A_t(Z)$.

Suppose τ is a random time and define $A_\tau(Z)$, the integral from 0 up to and including τ of Z with respect to A , by $A_\tau(z)(w) = A_{\tau(w)}(Z)(w)$, in particular $A_\tau(Z) = A_\infty(Z)$ on $(\tau = \infty)$.

2.5. <u>Proposition</u>. For τ a random time, $A_\tau(Z)$ is F-measurable. If τ is a stopping time and A and Z are adapted, then $A_\tau(Z)$ is F_τ-measurable.

<u>Proof</u>. For the first assertion we use the right-continuity of $A(Z)$ to write

$$A_\tau(Z) = \lim_{n \to \infty} \sum_{k=0}^{\infty} A_{\frac{k+1}{2^n}}(Z) \, 1_{\left(\frac{k}{2^n} \leq \tau < \frac{k+1}{2^n}\right)} + A_\infty(Z) \, 1_{(\tau=\infty)}$$

and the measurability of $A_\tau(Z)$ follows from Proposition 2.4(i). For the second assertion one checks directly that $A_\tau(Z)$ is constant on F_τ-atoms: if $w \underset{\tau}{\sim} w'$, then $\tau(w) = \tau(w')$ so the range of integration for $A_\tau(Z)(w)$ and $A_\tau(Z)(w')$ are the same. Furthermore, since A is adapted the measures induced by $s \to A_s(w)$ and $s \to A_s(w')$ are identical on $[0,\tau(w)] = [0,\tau(w')]$ and finally, since Z is adapted the integrands $s \to Z_s(w)$, $s \to Z_s(w')$ agree on the interval of integration.   ▌

Suppose now that P is a CCPE. Since for every t, $A_t(Z)$ is F-measurable and non-negative, the expectation $PA_t(Z)$ is well-defined The next result is then trivial and the proof is omitted.

2.6. <u>Proposition</u>. Let $Z \geq 0$ be measurable and locally uniformly bounded and let A be an increasing process. With respect to P, the stochastic integral $A(Z)$ has the following properties:

(i) $A_t(Z) < \infty$ P-a.s. simultaneously for all $t \geq 0$ if A is P-finite;

(ii) $PA_t(Z) < \infty$ for all $t \geq 0$ if A is P-locally integrable;

(iii) $PA_\infty(Z) < \infty$ if Z is uniformly bounded and A is P-integrable.

                                                                        ▌

We shall now extend the definition of the stochastic integral in two directions.

If Z is locally uniformly bounded but not necessarily non-negative, the ordinary integral

$$A_t(Z)(w) = \int_{(0,t]} dA_s(w) Z_s(w)$$

converges for all t,w such that $t < a(w)$, and is then equal to $A_t(Z^+) - A_t(Z^-)$, where Z^+, Z^- denotes the positive and negative part of Z respectively. Hence if P is a CCPE and A is P-finite, the stochastic integral $A_t(Z)$ is defined P-a.s. simultaneously for all t and $|A_t(Z)| < \infty$. If now A' is another P-finite increasing process, the same is true about $A'(Z)$, and so, almost surely, $|A_t(Z) - A'_t(Z)| < \infty$ simultaneously for all t. Introducing $D = A - A'$ we shall write

$$D_t(Z) = \int_{(0,t]} dD_s Z_s = A_t(Z) - A'_t(Z)$$

and call the almost surely defined random variable $D_t(Z)$ the stochastic integral over $(0,t]$ with respect to D. These considerations are important enough to be summarized as

2.7. <u>Proposition</u>. Let P be a CCPE, let A,A' be P-finite increasing processes and let Z be a real-valued, measurable, locally uniformly bounded process. Then P-almost surely the difference process $D = A-A'$ is defined and $|D_t| < \infty$ simultaneously for all t, and P-almost surely the stochastic integral $D(Z)$ is defined and $|D_t(Z)| < \infty$ simultaneously for all t. ∎

The extended stochastic integral just defined satisfies P-almost surely (i.e. for all paths w outside a fixed P-null set $N \in F$) properties (iii), (iv) (both for any $\alpha, \alpha' \in \mathbb{R}$) and, as already stated, property (vi) of Proposition 2.4. Of course also $D(Z)$ is right-con-

3.2.6

tinuous almost surely.

Property (ii) may be given the following form: if A,A' , hence
D , and Z are adapted (predictable), there is for every $t \geq 0$ a
set $N_t \in F_t$, $(N_t \in F_{t-})$ with $PN_t = 0$ such that $D_t(Z)(w)$ is defined and finite for all $w \notin N_t$ and the process $(D_t(Z)1_{N^c_t})_{t \geq 0}$ is
adapted (predictable). This is obvious since the integrals with respect
to D are defined in terms of differences of the integrals considered
in Proposition 2.4.

Finally, the monotone convergence discussed in Proposition 2.4(v)
may be replaced by a dominated convergence: if $Z \geq 0$ is locally uni-
formly bounded and the sequence (Z_n) satisfies $|Z_n| \leq Z$, then
P-a.s. $\lim_{n \to \infty} D_t(Z_n) = D_t(Z)$ simultaneously for all t if $Z_n \to Z$.

If P is such that $P\bar{N}_t < \infty$ for all t, then by Theorem 2.2.13,
$P\Lambda^y_t < \infty$ for all y,t . Proposition 2.7 therefore enables us to discuss stochastic integrals with respect to the martingales $M^y = N^y - \Lambda^y$.
The main properties of these integrals are given in the next result.

2.8. <u>Theorem</u>. Let P be a CCPE of class H^E with locally finite expectations: $P\bar{N}_t < \infty$ for all t, and let Z be a real-valued, predictable process.

(i) Suppose that $Z \geq 0$. Then

$$PN^y_t(Z) = P\Lambda^y_t(Z)$$

for every $y \in E$, $t \geq 0$.

(ii) Suppose that Z is locally uniformly bounded. Then the stochastic integral $M^y(Z)$ is well-defined P-almost surely, and $M^y(Z)$
is a P-martingale which is locally square integrable:
$M^y(Z) \in M^2_0(P)$.

(iii) Suppose that Z is locally uniformly bounded. Then

$$<M^y(Z)> = \Lambda^y(Z^2) \qquad\qquad (y \in E).$$

More generally, if Z' is another locally uniformly bounded, predictable process, then $M^y(Z)M^z(Z') \in M(P)$ if $y \neq z$ and

$$\langle M^y(Z), M^y(Z') \rangle = \Lambda^y(ZZ') \qquad (y \in E) ,$$

$$\langle M^y(Z), M^z(Z') \rangle = 0 \qquad (y \neq z \in E) .$$

Proof. Theorem 2.2.13 may be viewed as a special case of this result for $Z \equiv 1$, and the proof of the present theorem follows the same pattern as that of Theorem 2.2.13 and its one-dimensional analogue, Theorem 1.5.1. Therefore, it is basic that we show

$$(2.9) \qquad PN^y_{t \wedge \tau_1}(Z) = P\Lambda^y_{t \wedge \tau_1}(Z)$$

$$(2.10) \qquad P(M^y_{t \wedge \tau_1}(Z) M^y_{t \wedge \tau_1}(Z')) = P\Lambda^y_{t \wedge \tau_1}(ZZ') ,$$

$$(2.11) \qquad P(M^y_{t \wedge \tau_1}(Z) M^z_{t \wedge \tau_1}(Z')) = 0$$

for all $t \geq 0$, $y \neq z \in E$.

Of these three identities, we shall only prove the first. The two others follow by similar computations, cf. the last part of the proof of Theorem 1.5.1 and the proof of (2.2.14).

For the proof of (2.9) we need only know the behavior of Z on $[0, \tau_1]$, and find that since Z is predictable, there is a measurable function $f: [0, \infty) \to \mathbb{R}$, bounded on finite intervals, such that

$$Z_t = f(t) \qquad \text{on } (\tau_1 \geq t) ,$$

cf. the representation (1.13) of a predictable martingale. It is vital that in particular

$$Z_{\tau_1} = f(\tau_1) \qquad \text{on } (\tau_1 < \infty) ,$$

and it is for this the assumption that Z be predictable is needed. Now, by direct computation

$$PN_{t\wedge\tau_1}^y(Z) = P(Z_{\tau_1}; \tau_1 \leq t, Y_1 = y)$$

$$= \int_0^t ds\ \mu_0(s)\, e^{-\int_0^s \mu_0}\ f(s)\, \pi_0(s-,y),$$

$$P\Lambda_{t\wedge\tau_1}^y(Z) = P\left(\int_0^{t\wedge\tau_1} ds\ \lambda^y(s)\, f(s)\right)$$

$$= \int_0^t ds\ \mu_0(s)\, e^{-\int_0^s \mu_0} \int_0^s du\ \mu_0(u)\, \pi_0(u-,y)\, f(u)$$

$$+ \left(e^{-\int_0^t \mu_0}\right) \int_0^t ds\ \mu_0(s)\, \pi_0(s-,y)\, f(s),$$

and if f is right-continuous, both expressions are continuous in t and differentiable from the right, so (2.9) is verified by differentiation. Thus the space of functions f for which the two expressions above are equal contain all right-continuous step functions, and since the space is closed under the formation of monotone increasing limits, it contains all measurable functions, bounded on finite intervals.

Using the fundamental conditioning result, Theorem 2.1.22, to imitate the proof of Theorem 1.5.1, one finds that (2.9)-(2.11) remain valid when $t \wedge \tau_1$ is replaced by $t \wedge \tau_n$. But if $Z \geq 0$, the arguments in the proof of Theorem 1.5.1 involving monotone convergence, Fatou's lemma and optional sampling to deduce (1.5.2'), (1.5.3') from (1.5.2) and (1.5.3), yield

$$PN_t^y(Z) = P\Lambda_t^y(Z),\quad P(M_t^y(Z))^2 = P\Lambda_t^y(Z^2)$$

for all P, and from this (i), (ii) and the first part of (iii) follow for $Z \geq 0$ via another application of Theorem 2.1.22. The same kind of reasoning establishes the last parts of (iii), showing first that

$$PM_t^y(Z)\, M_t^y(Z') = P^y\Lambda(ZZ'),\quad PM_t^y(Z)\, M_t^z(Z') = 0$$

for $Z, Z' \geq 0$, $y \neq z$, and then using 2.1.22.

The assertions of the theorem for general locally uniformly bounded Z, Z' follow by splitting Z, Z' into their positive and negative parts and using the linearity of the stochastic integrals and Proposition 1.19. ❙

The final definition of this section is included merely to place in the present context an important concept from the general theory. The definition will not be used in the sequel.

2.12. <u>Definition</u>. A random time τ is <u>predictable</u> if $(\tau \leq t) \in F_{t-}$ for all $t \geq 0$. ❙

In particular a predictable random time is a stopping time. If τ is predictable, the right-continuous process $Z_t = 1_{(\tau \leq t)}$ is predictable, so that by (2.9)

$$(2.13) \qquad PN^Y_{\tau_1}(Z) = P\Lambda^Y_{\tau_1}(Z)$$

if $P \in H^E$ with $P\bar{N}_t < \infty$.

Clearly τ_1 is not predictable. Instead, try for τ predictable such that $\tau = \tau_1$ P-a.s. Then $Z_{\tau_1} = 1$ a.s. on $(\tau_1 < \infty)$, and we find $PN^Y_{\tau_1}(Z) = P(\tau_1 < \infty, \ Y_1 = y)$ while $\Lambda^Y_{\tau_1}(Z) = 0$ a.s. It follows that it is possible to find such a τ only for P the trivial CCPE P_0 with $P_0(\tau_1 = \infty) = 1$.

Notice that the process $Z_t = 1_{(\tau_1 < t)}$ is left-continuous and adapted, hence predictable, so that for this process (2.13) holds. Of course this can also be verified directly since now $Z_{\tau_1} = 0$ on $(\tau_1 < \infty)$.

It is customary to call a random tine τ predictable, if there is an increasing sequence (τ_n) of stopping times with $\lim \tau_n = \tau$ and such that $\tau_n < \tau$ everywhere on $(\tau > 0)$ for every n. We shall now show that such a time is predictable in the sense of Definition 2.12.

To see this, let $\tau(w) > t$, where $t > 0$, $w \underset{t-}{\sim} w'$. We must show that $\tau(w') > t$. But since $\tau_n \uparrow \tau$, also $\tau_n(w) > t$ for n

sufficiently large, $n \geq n_0$ say. Therefore $\tau_n(w) > t - \varepsilon$ for $n \geq n_0$ and all $0 < \varepsilon \leq t$ and since $w' \underset{t-\varepsilon}{\sim} w$ and τ_n is a stopping time we get $\tau_n(w') > t - \varepsilon$ for $n \geq n_0$ and all $0 < \varepsilon \leq t$, i.e. $\tau_n(w') \geq t$ for $n \geq n_0$. Since $\tau_n < \tau$ it follows that $\tau(w') > t$ as desired.

As an example of an interesting predictable stopping time, consider for $\gamma \geq 0$,

$$\tau = \inf\{t \geq 0 : \Lambda_t^y \geq \gamma\},$$

where Λ is either the integrated intensity for a process of class H^E, or the accumulated intensity for a process of class $\mathcal{D}^E$.

Notes.

The standard reference for the general theory of processes including martingales and stochastic integrals, is Dellacherie and Meyer (1975) and (1980). Two fine surveys of stochastic integration theory are Dellacherie (1980) and Shiryayev (1981). See also the introductury papers by Williams, Rogers and Elliott in Williams (ed.) (1981).

Definition 1.4 of a predictable process is suited to the canonical setup, but is different from the usual one. In the general theory, a process $Z = Z_t(\omega)$, defined on a filtered space satisfying the usual conditions, is predictable if as a function of t and ω, it is measurable with respect to the σ-algebra generated by the class of adapted, left-continuous processes.

The general theory comprises various results about representations of martingales. For jump processes such a representation is given in part I of Boel, Varaiya and Wong (1975), Chou and Meyer (1975), Jacod (1975), Davis (1976) and Liptser and Shiryayev (1977-78), Chapter 19. For canonical processes, a result of this type and its proof is indicated in Exercise 2 below. The formulation is very similar to Davis (1976).

Exercises.

1. The proof of Proposition 3.1.12 utilizes a representation of predictable processes. A similar result is valid for adapted process.
   Thus, let Z defined on W^E be adapted, and show that for every
   $n \geq 0$, $0 < t_1 < \cdots < t_n$, $y_1, \cdots, y_n \in E$, writing
   $z_n = (t_1, \cdots, t_n, y_1, \cdots, y_n)$, there exists a function
   $f_n(z_n; \cdot) : [t_n, \infty) \to \mathbb{R}$ such that

   $$Z_t = f_n(\xi_n; t) \qquad \text{on } (\tau_n \leq t < \tau_{n+1}) ,$$

   in particular $Z_{\tau_n} = f_n(\xi_n; \tau_n)$.　　　　　　　　　▌

2. Let P be a one-dimensional, stable CCP of class H with finite
   expectations locally, and let $m = (m_t)_{t \geq 0}$ be a P-martingale
   with $m_0 = 0$.

   A well-known result from the general process theory states that
   m may be represented as a stochastic integral

   $$(1) \qquad\qquad m_t = \int_{(0,t]} M(ds) Z_s$$

   of a predictable process Z with respect to the fundamental martingale $M = N-\Lambda$. The purpose of this exercise is to indicate
   how this representation theorem may be established.

   As an adapted process m may, according to Exercise 1, be written

   $$m_t = f_n(t_1, \cdots, t_n; t)$$

   on $(\tau_1 = t_1, \ldots, \tau_n = t_n, N_t = n)$ where for each $n \geq 0$, $t_1 < \cdots < t_n$,
   $f_n(t_1, \cdots, t_n; \cdot)$ is a function from $[t_n, \infty)$ to $\mathbb{R}$, writing of
   course $t_0 = 0$. In particular

   $$m_{\tau_n} = f_n(\tau_1, \cdots, \tau_n; \tau_n).$$

Similarly, as a predictable process Z may be written

$$Z_t = g_n(t_1, \cdots, t_n; t)$$

on $(\tau_1 = t_1, \cdots, \tau_n = t_n, N_{t-} = n)$ with $g_n(t_1, \cdots, t_n; \cdot)$ defined on (t_n, ∞), so that in particular

$$Z_{\tau_n} = g_{n-1}(\tau_1, \cdots, \tau_{n-1}; \tau_n),$$

cf. (3.1.13).

Given the martingale m, i.e. the functions f_n, the problem is to find Z, i.e. the g_n, such that (1) holds. (Since Λ, hence M is only determined up to P-indistinguishability, one can with a given version of Λ only hope to obtain (1) P-a.s., simultaneously for all t. But here we shall not worry much about exceptional sets).

Show that (1) is satisfied as an identity everywhere on W if and only if for every $n \geq 0$, $t_1 < \cdots < t_n \leq t$

$$(2) \qquad f_n(t) = \sum_{k=1}^{n} \left(g_{k-1}(t_k) - \int_{t_{k-1}}^{t_k} ds\, g_{k-1}(s)\mu_{k-1}(s) \right)$$

$$- \int_{t_n}^{t} ds\, g_n(s)\mu_n(s),$$

where

$$f_n(t) = f_n(t_1, \cdots, t_n; t),$$

$$g_k(s) = g_k(t_1, \cdots, t_k; s),$$

$$\mu_k(s) = \mu_{k, t_1, \cdots, t_k}(s).$$

It is obvious that this equation can only be solved if $f_n(t)$ is an absolutely continuous function of t. Why this is true will follow from a question below. Instead, assume for now, that f_n

is differentiable from the right, and that all $\mu_k(s) > 0$. Show that then necessarily

$$g_n(t) = - D^+ f_n(t) / \mu_n(t)$$

for all $n \geq 0$, $t > t_n$. Deduce from this that (2) can be solved if and only if

(3) $\quad f_n(t_n) = \sum_{k=1}^{n} \left(- D^+ f_{k-1}(t_k)/\mu(t_k) + f_{k-1}(t_k) - f_{k-1}(t_{k-1}) \right)$

for all $n \geq 0$, $t_1 < \cdots < t_n$.

To show that (3) holds, use that m is a martingale: show, using optional sampling, that a.s.

$$P(m_{t \wedge \tau_{n+1}} | F_{\tau_n}) 1_{(\tau_n \leq t)} = m_{\tau_n} 1_{(\tau_n \leq t)} ,$$

and then show that this may be rewritten as, with $\xi_n = (\tau_1, \cdots, \tau_n)$,

$$f_n(\xi_n;t) \exp\left(-\int_{\tau_n}^{t} ds \, \mu_{n\xi_n}(s) \right)$$

$$+ \int_{\tau_n}^{t} ds \, \mu_{n\xi_n}(s) \exp\left(-\int_{\tau_n}^{s} \mu_{n\xi_n} \right) f_{n+1}(\xi_n, s; s) = f_n(\xi_n; \tau_n)$$

a.s. on $(\tau_n \leq t)$. Show from this that for almost all values of $\tau_1, \cdots, \tau_n$, the function $t \to f_n(\xi_n;t)$ is absolutely continuous (cf. the remark after (2) above), and show also that (3) holds, at least for almost all possible values of $t_1, \cdots, t_n$.

Finally, discuss briefly how the proof may be modified if it is not assumed that all $\mu_k(s) > 0$. (Then exceptional sets enter already in (2)).

The representation (1) generalizes to the multivariate case provided one replaces M by the vector $(M^y)_{y \in E}$ and Z by a vector $(Z^y)_{y \in E}$ of predictable processes.

3. Let P be a CCPE of class H^E with $P\bar{N}_t < \infty$ for all t. By the uniqueness assertions in Theorem 3.1.9 and Proposition 3.1.18, the integrated intensity $\Lambda = (\Lambda^y)_{y \in E}$ is the unique predictable, componentwise increasing process with $\Lambda_0^y = 0$, such that $M^y = N^y - \Lambda^y$ for $y \in E$ are orthogonal martingales.

Use this characterization of the integrated intensity to give an alternative proof of Theorem 2.3.1.   ∎

4. Show that the stochastic integral $A(Z)$ given by (3.2.2) is a predictable process if Z is adapted and A is continuous. ∎

5. Let Z, Z' be predictable processes as in Theorem 3.2.8. Then there exists functions f, f' such that

$$Z_t = f(t), \quad Z_t' = f'(t) \quad \text{on} \quad (\tau_1 \geq t).$$

Now write $\mu(s) = \mu_0(s)$, $\pi(s) = \pi_0(s-, y)$ and show that

$$P(M_{t \wedge \tau_1}^y (Z) M_{t \wedge \tau_1}^y (Z'))$$

$$= e^{-\int_0^t \mu} \left(\int_0^t \mu \pi f \right) \left(\int_0^t \mu \pi f' \right) + \int_0^t ds \, \mu(s) e^{-\int_0^s \mu} \pi \left(f(s) - \int_0^s \mu \pi f \right) \left(f'(s) - \int_0^s \mu \pi f' \right)$$

$$+ \int_0^t ds \, \mu(s) e^{-\int_0^s \mu} (1 - \pi) \left(\int_0^s \mu \pi f \right) \left(\int_0^s \mu \pi f' \right).$$

Show also that

$$P \, \Lambda_{t \wedge \tau_1}^y (ZZ') = e^{-\int_0^t \mu} \int_0^t \mu \pi f f' + \int_0^t ds \, \mu(s) e^{-\int_0^s \mu} \int_0^s \mu \pi f f'.$$

Deduce (3.2.10) from this, first for nice f, f' and then by approximation for general f, f'.

Give a similar derivation of (3.2.11).   ∎

6. Show that the process Λ in Exercise 1.9 is predictable.

4. THE MULTIPLICATIVE INTENSITY MODEL.

4.1. Definition of the full Aalen model.

We shall discuss statistical models for counting processes, i.e. families of CCPE's of class H^E, for which the intensity has the form

$$(1.1) \qquad \lambda_{t-}^y = \alpha^y(t-)z_{t-}^y ,$$

where for each y, α^y is a non-negative function defined on $[0,\infty)$ and z^y is a process defined on W^E. Here z^y is to be thought of as something which is observed, while the α^y are the unknown parameters.

We know from Proposition 2.2.2 which conditions a process λ_- must satisfy in order that it be the intensity process of a CCPE. These conditions lead to some natural conditions on the α^y and z^y, which together with others required to give the CCPE's under consideration additional desirable properties are summarized below as the

Basic assumptions. It is assumed that each $\alpha^y \in A$, where A is the space of functions $\alpha: [0,\infty) \to [0,\infty]$ which are right-continuous with left-limits and satisfy $\int_0^t ds\alpha(s) < \infty$ for all t, i.e. α is the intensity function for a probability on $(0,\infty]$ with smooth density and termination point ∞.

It is assumed that $z = (z^y)_{y\in E}$ belongs to the space z^E of a-dapted processes for which each component is non-negative with right-continuous, left-limit paths such that for every t

$$(1.2) \qquad \overline{z}_t \le a + b\overline{N}_t ,$$

where $a \ge 0$, $b \ge 0$ are constants and $\overline{z} = \sum_y z^y$, and such that for every $y \in E$ the process

$$(1.3) \qquad \frac{1}{z_-^y} 1_{(z_-^y > 0)}$$

is locally uniformly bounded, where z_-^y is the left regularization of

Z^y taking the value Z^y_{t-} at time t .   I

As we shall see in a moment, the boundedness condition (1.2) ensures that $P\bar{N}_t < \infty$ for all t if P has intensity (1.1). The process in (1.3) at time t takes the value

$$\frac{1}{Z^y_{t-}} \ 1_{(Z^y_{t-} > 0)}(t)$$

and the local boundedness of this is required for the estimation theory.

The conditions on the α^y may be relaxed if one is apriori interested in studying the process on a given time interval $[0,t_0]$. It
is then sufficient that the termination points for the probabilities with intensity α^y be $> t_0$.

1.4. <u>Proposition</u>. If $\alpha = (\alpha^y)$ and $Z = (Z^y)$ satisfy the basic assumptions, then (1.1) defines the intensity for a CCPE P_α such that
$P_\alpha \bar{N}_t < \infty$ for all t .

<u>Proof.</u> Writing $\gamma(t) = \max_y \alpha^y(t)$ we have (working now with the right-
continuous regularization of λ_-)

$$\bar{\lambda}_t \leq \gamma(t) \ (a + b\bar{N}_t) ,$$

so by a stochastic comparison   argument similar to the one used in the
proof of Proposition 1.4.5,the result will follow if we show that on the
(unstable) one-dimensional path-space $\bar{W}$, the CCP $\bar{P}$ with (right-continuous) intensity

$$\gamma(t) \ (a + bN_t)$$

(where N_t refers to $\bar{W}$) satisfies $\bar{P}N_t < \infty$ for all t .

If a = 0 this is evident since then G_0 has intensity 0 and
$\bar{P}(N_t = 0) = 1$, so we shall assume that a > 0 . Since the statement
if proved for one function γ , remains true for all functions $\leq \gamma$ by
stochastic domination, we may and shall assume that γ satisfies
$\int_0^\infty ds\gamma(s) = \infty$.

Let $\Gamma(t) = \int_0^t ds\, \gamma(s)$ which is finite by the basic assumptions, and let Γ^{-1} be the right-continuous inverse of Γ. By Proposition 1.4.8(c), $\int_0^{\tau_n} ds\, \gamma(s)\,(a+bN_s) < \infty$ $\bar{P}$-a.s. for all n. Since we are assuming $\int_0^\infty \gamma = \infty$ and $a > 0$, this forces $\tau_n < \infty$ $\bar{P}$-a.s. But then, also by Proposition 1.4.8(a), $\int_{\tau_n}^{\tau_{n+1}} ds\, \gamma(s)\,(a+bN_s) > 0$ whence $\Gamma(\tau_n) < \Gamma(\tau_{n+1})$ $\bar{P}$-a.s. for every n. If therefore N^* is given by $N_u^* = N_{\Gamma^{-1}(u)}$ for $u \geq 0$, then $\bar{P}$-a.s. the process N^* has infinitely many jumps all of size 1, and $\tau_n^* = \Gamma(\tau_n)$ with τ_n^* denoting the time of the n'th jump of N^* (cf. the beginning of the proof of Proposition 1.4.6). Now

$$
\begin{aligned}
\bar{P}(\tau_{n+1}^* - \tau_n^* > v \,|\, F_{\tau_n}) &= \bar{P}\left(\int_{\tau_n}^{\tau_{n+1}} ds\, \gamma(s) > v \,|\, F_{\tau_n}\right) \\
&= \bar{P}\left(\int_0^{\tau_{n+1}-\tau_n} ds\, \gamma(s+\tau_n)\,(a+bn) > v(a+bn) \,|\, F_{\tau_n}\right) \\
&= e^{-v(a+bn)}
\end{aligned}
$$

by Proposition 1.1.4 and the fact that $s \to \gamma(s+\tau_n)\,(a+bn)$ is the intensity for the conditional distribution of $\tau_{n+1} - \tau_n$ given F_{τ_n}. Thus for N^* the waiting times between jumps are stochastically independent, the waiting time in state n being exponential with intensity $a+bn$. But it is well known, that this linear growth process has finite expectations so that $\bar{P}N_u^* = \bar{P}N_{\Gamma^{-1}(u)} < \infty$ for all u. Since $\lim_{u\uparrow+\infty} \Gamma^{-1}(u) = \infty$ it follows that $\bar{P}N_t < \infty$ for all t. ∎

1.5. <u>Definition</u>. The <u>full Aalen model</u> for a given process $Z \in Z^E$ is the family

$$P(Z) = \{P_\alpha : \alpha = (\alpha^y)_{y\in E} \in A^E\}$$

where P_α is the CCPE with intensity $\lambda_- = (\lambda_-^y)_{y\in E}$ given by

$$\lambda_{t-}^y = \alpha^y(t-)\,z_{t-}^y \,.$$ ∎

We again emphasize that by Proposition 1.4, every P_α is well-defined as a CCPE and satisfies $P_\alpha \bar{N}_t < \infty$ for all $t \geq 0$.

The basic statistical problem in the Aalen model is to estimate α and discuss asymptotic distibutional properties of the estimator.

The full Aalen model is non-parametric in the sense that each component α^y of the unknown α is allowed to range freely in the class A. By restricting the possible choices for the α^y one may obtain smaller non-parametric models or even parametric ones, or mixtures of non-parametric and parametric models, see Section 4.5 below.

We conclude this section with some examples that illustrate the variety of cases covered by the Aalen setup.

1.6. Example. Let $X_1, \cdots, X_r$ be i.i.d. strictly positive random variables with survivor function G having intensity $\mu \in A$. As discussed in Example 1.2.6 , $K_t = \sum_{i=1}^{r} 1_{(X_i \leq t)}$ defines a counting process, and from Example 1.3.8 we find that for the corresponding CCP, P_μ, the intensity for the distribution of τ_{n+1} given $\tau_1 = t_1, \cdots, \tau_n = t_n$ is $\mu_{nt_1 \cdots t_n}(t) = \mu(t)(r-n)$ for $n \leq r$, $t > t_n$, so that the intensity for P_μ is

$$\lambda_{t-} = \mu(t-)(r - N_{t-})^+ .$$

Thus $P(Z) = \{P_\mu : \mu \in A\}$ is the full Aalen model for $Z_t = (r-N_t)^+$. (Clearly this Z satisfies the basic assumptions!)

We might instead have considered the multivariate counting process $K = (K^i)_{i \in E}$ with $K_t^i = 1_{(X_i \leq t)}$ and $E = \{1, \cdots, r\}$. The corresponding CCPE, P_μ^E, has intensity $\lambda_- = (\lambda_-^i)_{i \in E}$ where

$$\lambda_{t-}^i = \mu(t-)(1 - N_{t-}^i)^+ = \mu(t-)1_{(\tau_1^i \geq t)} .$$

Thus $P = \{P_\mu^E : \mu \in A\}$ is a submodel of the full Aalen model with $Z_t^i = (1 - N_t^i)^+$ and $\alpha^i = \mu$ for all i. (Of course the full Aalen model for $Z = (Z^i)$ arises when the X_i are assumed to be independent but with different distributions, each having intensity in A). ∎

1.7. Example. Suppose as in Examples 1.2.7 and 2.1.6 that we are given not only the X_i, but also censoring variables U_i. As announced at the end of Example 2.1.6, we shall work conditionally on the U_i taking given values u_i, and assume that the X_i are i.i.d. and independent of the U_i. (To make another point we shall however return briefly to the case of general X_i and U_i in the next example). Introducing for $i = 1, \cdots, r$

$$K_t^i = 1_{(X_i \leq t \wedge u_i)} ,$$

we have that the process K from Example 1.2.7 counting the observed number of deaths, is $K = \Sigma K^i$.

If the common survivor function for the distributions of the X_i is assumed to have intensity $\mu \in A$, it is seen that the intensity for the CCPE, with $E = \{1, \cdots, r\}$ generated by $(K^i)_{1 \leq i \leq r}$, is

$$\lambda_t^i = \mu(t-) \, 1_{(\tau_1^i \wedge u_i \geq t)} .$$

The indicator function is (of course) left-continuous. It is 1 if individual i is under observation or at risk (i.e. not dead and not censored) just before t.

For μ varying freely in A we obtain a submodel of the full Aalen model with $Z = (Z^i)$, $Z_t^i = 1_{(\tau_1^i \wedge u_i > t)}$. |

1.8. Example. As in the previous example, suppose given lifetimes $X_1, \cdots, X_r$ and censoring times $U_1, \cdots, U_r$ but for the time being, do not make any assumptions about their joint distribution.

For $i = 1, \cdots, r$, consider the process K^i defined in Example 1.7, and also introduce (cf. Example 2.1.6)

$$K_t^{*i} = 1_{(U_i \leq t \wedge X_i)} .$$

Assuming now that $\mathbb{P}(X_i = U_i) = 0$, (K^i, K^{*i}) is a two-dimensionsional counting process, and assuming further that the distribution of $X_i \wedge U_i$ has smooth density with intensity γ_i and writing

$$p_i(t) = \mathbb{P}(X_i \le U_i | X_i \wedge U_i = t)$$

$$= \mathbb{P}(X_i < U_i | X_i \wedge U_i = t),$$

it is found that (K^i, K^{*i}) generates a two-dimensional CCPE, provided p_i is right-continuous with left-limits. The intensity process becomes $(\lambda_-^i, \lambda_-^{*i})$, where

$$\lambda_{t-}^i = \gamma_i(t-)p_i(t-)1_{(\tau_1 \ge t)} = \gamma_i(t-)p_i(t-)(1-\bar{N}_{t-})^+,$$

$$\lambda_t^{*i} = \gamma_i(t-)(1-p_i(t-))1_{(\tau_1 \ge t)} = \gamma_i(t-)(1-p_i(t-))(1-\bar{N}_t)^+,$$

which for γ_i, p_i varying freely, gives the full Aalen model for $(z^i, z^{*i}) = ((1-\bar{N})^+, (1-\bar{N})^+)$.

The model has been derived solely from assumptions on the distribution of $X_i \wedge U_i$, and of course does not contain much information about the joint behavior of X_i and U_i: as we shall see in a moment, the same model may be derived by assuming X_i and U_i to be independent.

In the model X_i and U_i enter in a completely symmetric fashion, and observation of (K^i, K^{*i}) amounts to observing $X_i \wedge U_i$. But if X_i and U_i are thought of as two possible causes of death, we then have the standard scheme for discussing (two) <u>competing risks</u>.

That the same model arises with independent X_i, U_i is easily seen: that the distribution of X_i, (U_i) has intensity α_i, (β_i) with X_i, U_i independent, corresponds to taking

$$\gamma_i(t) = \alpha_i(t) + \beta_i(t), \quad p_i(t) = \alpha_i(t)/(\alpha_i(t) + \beta_i(t)). \quad \blacksquare$$

1.9. <u>Example</u>. Consider a Markov chain (X_t) on the finite state-space S as in Example 2.1.19, and let $K_{ij}(t)$ denote the number of transitions from i to j for X on $[0,t]$. With the notation from the example, assume that $G^{(i)}$ has smooth density with intensity μ_i, and that the jump probabilities $\pi_{ij}(t-)$ are left-continuous with right-limits. With $E = \{(i,j) \in S^2 : i \neq j\}$ and P^i the CCPE generated by $K = (K_{ij})$ when $X_0 = i$ a.s., it is seen that the intensity $\lambda_- = (\lambda_-^{ij})$ for P^i is given by

$$\lambda_{t-}^{ij} = \mu_i(t-)\pi_{ij}(t-) 1_{(J_{\overline{N}_{t-}} = i)} \, ,$$

writing $Y_n = (I_n, J_n)$ as in Example 2.1.11. When the μ_i and π_{ij} are allowed to range freely, this gives a full Aalen model with $Z = (Z^{ij})_{((i,j) \in E)}$ determined by

$$z_t^{ij} = 1_{(J_{\overline{N}_t} = i)} \, .$$

It is convenient to use the <u>transition intensities</u> $\alpha_{ij}(t) = \mu_i(t)\pi_{ij}(t)$ as parameters, rather than the μ_i and π_{ij}. Of course each α_{ij} is an arbitrary member of A. ▐

1.10. <u>Example</u>. If in Example 1.3.6 about renewal sequences, the survivor function G has smooth density with intensity μ, the intensity for the CCP generated by the renewal sequence becomes

$$\lambda_{t-} = \mu((t-\tau_{N_{t-}})-) .$$

With μ varying freely in A, this gives a model, which is <u>not</u> a multiplicative intensity model. Some more general Markov renewal models have been treated by Gill (1980). ▐

4.2. Product models and sufficient reductions.

The most important statistical models for counting processes a-
rise, when one considers several independent counting processes, having one or more parameters in common.

Here the framework for discussing such models is provided by the
product processes introduced in Section 2.3. Recall from there that if
for $i = 1, \cdots, r$, P_i is a $CCPE_i$ of class H^{E_i}, then the product
$P = P_1 \otimes \cdots \otimes P_r$ may be viewed as a CCPE* of class H^{E^*} with $E^* =$
$\underset{i}{\cup} \{i\} \times E_i$. Recall also from Theorem 2.3.1 that the intensity for P
is determined in a simple manner from the intensities for each P_i.

We shall now assume that $E_i = E$, $1 \leq i \leq r$ and denote E^*,
which is the disjoint union of r copies of E, by rE. A typical
element of rE will be written (i,y) where $1 \leq i \leq r, y \in E$. As
in Section 2.3 we write $N = (N^{i,y})_{1 \leq i \leq r, y \in E}$ for the counting process
on W^{rE} and its components and $N^i = (N^{i,y})_{y \in E}$ for the i'th subpro-
cess. Also we shall write $\bar{N}^i = \underset{y}{\Sigma} N^{i,y}$, $\bar{N} = \underset{i}{\Sigma} \bar{N}^i$ and as usual denote
the jump times for N by τ_n, and the components jumping by Y_n.
For the intensity or integrated intensity we copy the notation used
for N, $N^{i,y}$, N^i, $\bar{N}^i$, $\bar{N}$. (Thus for instance $\lambda_-^{i,y}$ stands for the
(i,y)'th component of the intensity λ_-, and as such is defined on
W^{rE} in contrast with the notation from Section 2.3, where $\lambda_-^{i,y}$ de-
noted the y'th component of the intensity for the i'th process, i.e.
it was defined on W^E. Theorem 2.3.1 allows this switch in notation).

Let now $Z = (Z^{i,y})_{1 \leq i \leq r, y \in E}$ be a process on W^{rE} satisfying
the basic assumptions (1.2), (1.3), and such that for every i, $Z^i =$
$(Z^{i,y})_{y \in E}$ is a function of N^i alone.

2.1. Definition. The product Aalen model for Z is the family

$$P(Z) = \{ \underset{i=1}{\overset{r}{\otimes}} P_{i,\alpha} : \alpha = (\alpha^y)_{y \in E} \in A^E \}$$

of processes with type-set rE, where $P_{i,\alpha}$ is the CCPE with intensity $\lambda_-^i = (\lambda_-^{i,y})_{y \in E}$ given by

$$\lambda_-^{i,y} = \alpha^y(t-)\, z_{t-}^{i,y} \;. \qquad\qquad\blacksquare$$

In analogy with the previous notation, we shall write $\bar{z}^i = \sum_y z^{i,y}$, $\bar{z} = \sum_i \bar{z}^i$. The assumption above that z^i be a function of N^i is of course required for the product probabilities $\otimes P_{i,\alpha}$ to make any sense.

2.2. <u>Example</u>. The model for i.i.d. lifetimes from Example 1.6 appears now (with E a one-point set) as the product Aalen model for $Z = (z^i)_{1 \le i \le r}$, where $z^i = (1-N^i)^+$.

Similarly the model for i.i.d. lifetimes with censoring from Example 1.7, is the product Aalen Model for $Z = (z^i)_{1 \le i \le r}$, where

$$z_t^i = 1_{(\tau_1^i \wedge u_i > t)} \;. \qquad\qquad\blacksquare$$

Before formulating the main result about product models, we need a little more notation, and one new concept.

For $y \in E$, write $\tilde{N}^y = \sum_{i=1}^r N^{i,y}$, $\tilde{z}^y = \sum_{i=1}^r z^{i,y}$. Also, write ${}^0N_t$, ${}^0F_t$ etc. for the counting process and filtration etc. on the path-space W^E, (the symbols N_t, F_t etc. being reserved of course for the corresponding objects on W^{rE}).

Consider now a family $\Psi = (\psi_t^y)_{t \ge 0,\, y \in E}$ of mappings $\psi_t^y \colon W^E \to \mathbb{R}$. We shall say that Ψ is <u>linear</u> if

(i)   for every $t \ge 0$, $y \in E$, ψ_t^y is ${}^0F_t$-measurable;

(ii)   for every $y \in E$, ${}^0w \in W^E$ the mapping $t \to \psi_t^y({}^0w)$ is right-continuous with left-limits;

(iii)   for every $t \ge 0$, $y \in E$, $\sum_{i=1}^r \psi_t^y \circ (N^{i,z})_{z \in E} = \psi_t^y \circ (\tilde{N}^z)_{z \in E}$.

2.3. <u>Theorem</u>. (a) Given the product Aalen model $P(Z)$, consider for any $t \ge 0$ the model $P^t(Z)$ consisting of the restriction to F_t of each $P \in P(Z)$. A sufficient statistic for the model $P^t(Z)$ is then given by

$$\left(\widetilde{N}_s^y , \widetilde{Z}_s^y \right)_{s \leq t, y \in E} .$$

(b)   If in particular there exists a linear family $\Psi = (\psi_t^y)$ of mappings $\psi_t^y \colon W^E \to \mathbb{R}$ and there is for every i,y a right-continuous, left-limit function $a^{i,y} \colon [0,\infty) \to \mathbb{R}$ such that for every s,i,y

$$(2.4) \qquad z_s^{i,y} = a^{i,y}(s) + \psi_s^y \circ (N^{i,z})_{z \in E}$$

P-a.s. for all $P \in P(Z)$, then $(\widetilde{N}_s^y)_{s \leq t, y \in E}$ is sufficient for $P^t(Z)$. Moreover, the canonical counting process with type-set E generated by $\widetilde{N} = (\widetilde{N}^y)_{y \in E}$ under $P = \otimes P_{i,\alpha} \in P(Z)$, has intensity ${}^o\lambda_- = ({}^o\lambda_-^y)_{y \in E}$, where

$$(2.5) \qquad {}^o\lambda_{t-}^y = \alpha^y(t-) \, {}^oz_{t-}^y ,$$

writing

$$ {}^oz_t^y = \left(\sum_{i=1}^r a^{i,y}(t) + \psi_t^y \circ ({}^oN^z)_{z \in E} \right)^+ .$$

Finally, if ${}^oz = ({}^oz^y)_{y \in E}$ satisfies the basic assumptions, the model of CCPE's determined this way, is the full Aalen model for ${}^oz$.

<u>Proof.</u> (a) Consider for every i, a Poisson probability Π_{μ_i} on W^E, where $\mu_i = (\mu_i^y)_{y \in E}$. If $\Pi_\mu = \otimes \Pi_{\mu_i}$, $P = \otimes P_{i,\alpha}$ the likelihood function for $P^t(Z)$ becomes, according to Theorem 2.4.1

$$L_t = \frac{dP^t}{d\Pi_\mu^t} = \prod_{i=1}^r \frac{dP_{i,\alpha}^t}{d\Pi_{\mu_i}^t} ,$$

which is proportional to

$$\prod_{i=1}^r \left[\exp\left(- \sum_y \int_0^t ds \; \alpha^y(s) z_s^{i,y} \right) \prod_{k=1}^{\overline{N}_t^i} \alpha^{y_k^i}(\tau_k^i -) \, z^{i,y_k^i}(\tau_k^i -) \right],$$

where τ_k^i is the time of the k'th jump of the i'th subprocess, and y_k^i is the component jumping at that time. But here the factors involving α may be written

$$(2.6) \qquad \exp\left(-\sum_y \int_0^t ds\ \alpha^y(s)\, \tilde{z}_s^y\right) \prod_{y \in E} \prod_{k=1}^{\tilde{N}_t^y} \alpha^y(\tilde{\tau}_k^y-) ,$$

with $\tilde{\tau}_k^y$ the time of the k'th jump for $\tilde{N}^y$. Assertion (a) now follows from Neyman's criterion.

(b) Consider the identity

$$z_s^{i,y} = a^{i,y}(s) + \psi_s^y \circ (N^{i,z})_{z \in E}$$

and let for $t \geq 0$, C_t denote the set of paths for which it fails for some $s \leq t$ and some i,y . By assumption, given any $P \in P(Z)$, the identity holds P-a.s. for s,i,y fixed. But since everything is right-continuous in s , it follows that $P(C_t) = 0$. Since the $z^{i,y},\ N^{i,z}$ are adapted and ψ_s^y is ^oF_s-measurable, also $C_t \in F_t$. But then L_t may be replaced by $L_t' = L_t\, 1_{C_t^c}$ to give another density of P^t with respect to Π_μ^t , because for $F \in F_t$

$$P(F) = P(FC_t^c) = \Pi_\mu^t(L_t;FC_t^c) = \Pi_\mu^t(L_t';FC_t^c) = \Pi_\mu^t(L_t';F) .$$

Therefore the critical part (2.6) of the likelihood may be replaced by

$$1_{C_t^c}\ \exp\left(-\sum_y \int_0^t ds\ \alpha^y(s)\left(\sum_{i=1}^r a^{i,y}(s) + \psi_s^y \circ (\tilde{N}^z)_{z \in E}\right)\right) \prod_{y \in E} \prod_{k=1}^{\tilde{N}_t^y} \alpha^y(\tilde{\tau}_k^y-) ,$$

using the linearity of Ψ , and since by its definition, C_t does not depend on P (or α), Neyman's criterion shows that $(\tilde{N}_s)_{s \leq t}$ is a sufficient statistic.

By Proposition 2.2.2, to prove (2.5) it is sufficient to show that

$$(2.7) \qquad \lim_{h \downarrow \downarrow 0} \frac{1}{h}\, P(\bar{N}_{t+h} - \bar{N}_t \geq 1,\ \tilde{Y}_{t,1} = y \mid F_t) = {}^o\lambda_t^y \circ T$$

P-a.s. for every t . (Notice that $\bar{N} = \sum_y \tilde{N}^y$. Of course $\tilde{Y}_{t,1}$ is the component of $\tilde{N}$ jumping first after time t , and $T: W^{rE} \to W^E$ is the transformation given by $^oN_t^y \circ T = \tilde{N}_t^y$ for all t,y).

But for any $w \in W^{rE}$ and $h > 0$ so small that w does not jump on $(t, t+h)$, the conditional probability from (2.7) becomes

$$P(\tau_{t,1} \le t+h, \; Y_{t,1} = (i,y) \; \text{ for some } i \mid F_t)(w)$$

$$= \int_t^{t+h} ds \; \sum_{i=1}^{r} \lambda_{s-}^{i,Y}(w) \; \exp\left(-\int_t^s \bar{\lambda}_-(w)\right) \; .$$

Dividing by h and letting $h \!\downarrow\! 0$ gives

$$\lim_{h \downarrow 0} \frac{1}{h} \, P(\bar{N}_{t+h} - \bar{N}_t \ge 1, \; \tilde{Y}_{t,1} = y \mid F_t)(w) = \sum_{i=1}^{r} \lambda_t^{i,Y}(w) \, .$$

Thus the proof will be complete if we show that

$$\alpha^Y(t)\left(\sum_i a^{i,Y}(t) + \psi_t^Y \circ (\tilde{N}^z)_{z \in E}\right)^+ = \sum_i \lambda_t^{i,Y}$$

P-a.s. But from (2.4)

$$\alpha^Y(t)\left(\sum_i a^{i,Y}(t) + \psi_t^Y \circ (\tilde{N}^z)_{z \in E}\right) = \sum_i \lambda_t^{i,Y}$$

P-a.s., and since all $\lambda_t^{i,Y}$ and α^Y are ≥ 0 as intensities, we can safely take the positive part of the terms in brackets, and still keep an identity valid P-a.s.

The last part of (b) is trivial, and the proof is therefore complete. ∎

The theorem deserves some comments. Part (a) shows that when observing independent counting processes from a product Aalen model on a finite timeinterval $[0, t]$, there is always a sufficient reduction. Part (b) shows that in special cases, there is a particularly simple sufficient statistic, and that the original model may then be replaced by a much simpler one (consisting of one process with type-set E, rather than independent replicas of several).

The critical assumption in part (b) is (2.4). For applications it is quite important (see Example 2.8 below), that it is not assumed that the $a^{i,Y}$ or ψ_s^Y be ≥ 0. The point is that the conclusion in

(b) holds, provided (2.4) is an exact identity on the relevant part of W^{rE} (where of course the right-hand side is positive), while it does not matter whether strictly negative values of the expression on the right occur outside that part, as long as the linear structure is pre-served.

Because (2.4) holds only almost surely, it is not certain that ^{o}Z will satisfy the basic assumptions. Hence the condition for the last assertion in (b). It should be fairly clear, that the condition is not a genuine restriction.

Some examples of the linear Ψ used in the theorem:

$$\psi_t^y \circ (^oN^z)_{z \in E} = \begin{cases} \sum\limits_{z \in E} b^{yz}(t) \, ^oN_t^z \, , \\[2ex] \sum\limits_{z \in E} \int\limits_{(0,t]} \nu(ds) b^{yz}(s) \, ^oN_s^z \, , \\[2ex] \sum\limits_{z \in E} \int\limits_{(0,t]} \, ^oN^z(ds) b^{yz}(s) \end{cases}$$

with the b^{yz} suitable non-random functions and ν a suitable non-random measure. The first example is a special case of the second with $\nu = \varepsilon_t$, unit mass at t. The third example was considered by Aalen (1975).

2.8. **Example.** The product model $P(Z)$ from Example 2.2 with $z^i = (1-N^i)^+$ obviously satisfies that $z_t^i = 1-N_t^i$ P-a.s. for all $P \in P(Z)$, since with probability 1 no N^i jumps twice. Thus (2.4) holds with $\psi_t \circ {}^oN = -{}^oN_t$ (which may be < 0), hence $(\tilde{N}_s)_{s \leq t}$ is sufficient for $P^t(Z)$ and $\tilde{N}$ generates a one-dimensional canonical process with intensity

$$^o\lambda_t = \mu(t-)\left(\sum_{i=1}^{r} 1 - {}^oN_{t-}\right)^+ = \mu(t-)(r - {}^oN_{t-})^+ \, .$$

Of course we knew this already from Example 1.6. ∎

4.3. Estimation in the Aalen model.

Let $Z = (Z^{i,y})_{1 \leq i \leq r, y \in E}$ be as in Section 4.2, i.e. $Z \in Z^{rE}$ and for every i, Z^i is a function of N. Consider the product Aalen model $P(Z)$ as defined in Definition 2.1 and write $P_\alpha = \otimes P_{i,\alpha}$ for the member of $P(Z)$ with parameter α. Our aim is to estimate $\alpha = (\alpha^y)_{y \in E}$.

Since the intensity $\lambda_- = (\lambda_-^{i,y})$ for P_α is $\lambda_{t-}^{i,y} = \alpha^y(t-)z_{t-}^{i,y}$, it is clear that an observation w of the product process, which results in particular in a special value $z^{i,y}(w)$ for each $z^{i,y}$, only contains information about $\alpha^y(t-)$ for those values of t for which $z_{t-}^{i,y}(w) > 0$ for at least one i, i.e. the values of t with $\tilde{z}_{t-}^y(w) > 0$. Hence the estimator below only estimates the function $\alpha^y(t-)$ at such t-values.

Now introduce the process $\beta^* = (\beta^{*y})_{y \in E}$ by

$$\beta_t^{*y} = \int_0^t ds\ \alpha^y(s-)1_{(\tilde{z}_{s-}^y > 0)}(s)$$

$$= \int_0^t ds\ \alpha^y(s)1_{(\tilde{z}_s^y > 0)}(s) \ .$$

So given t, β_t^{*y} is the integral of the parameter $\alpha^y(s-)$ over that part of $[0,t]$, where there is any hope of estimating that parameter. We shall call β^{*y} an integrated intensity.

3.1. **Definition.** For every $t > 0$, the Aalen estimator in the model $P(Z)$ observed on the time interval $[0,t]$ of the integrated intensity β^{*y}, is the process $(\hat{\beta}_s^y)_{s \leq t}$, where

(3.2)
$$\hat{\beta}_s^y = \int_{(0,s]} \tilde{N}^y(du) \frac{1}{\tilde{z}_{u-}^y} 1_{(\tilde{z}_{u-}^y > 0)}(u) \ . \qquad \blacksquare$$

By the basic assumptions each process $(Z_-^{i,Y})^{-1} 1_{(Z_-^i, Y > 0)}$ is locally uniformly bounded, and the same is therefore true about $(\tilde{Z}_-^Y)^{-1} 1_{(\tilde{Z}_-^Y > 0)}$. Thus, the stochastic integral in (3.2) is well defined.

For the full Aalen model from Definition 1.5, the estimator of β^{*Y} becomes

$$\hat{\beta}_t^Y = \int_{(0,t]} N^Y(ds) \frac{1}{Z_{s-}^Y} 1_{(Z_{s-}^Y > 0)}(s) .$$

It must be stressed that the Aalen estimator is something which is defined, and not derived from general principles of statistical inference. That it is a reasonable estimator will appear from Theorem 3.3 below, and above all from the asymptotic theory of Chapter 5. One kind of reasoning leading to the Aalen estimator is the following: for small $h > 0$

$$\frac{1}{h} P_\alpha(\tilde{N}_t^Y - \tilde{N}_{t-h}^Y \mid F_{t-h}) \sim \alpha^Y(t-)\tilde{Z}_{t-}^Y .$$

Replacing the conditional expectation by the observation $\Delta\tilde{N}_t^Y$ of whether $\tilde{N}^Y$ jumps at t, suggests $\Delta\tilde{N}_t/\tilde{Z}_{t-}^Y$ as an estimator of $\alpha^Y(t-)$, and from this the Aalen estimator arises by integration.

Apart from being good in a statistical sense, the Aalen estimator has the major advantage, that it is extremely simple to calculate and to use.

There is one immediate and important problem in connection with Definition 3.1, namely how to interpret the Aalen estimator. No matter which value of α^Y is the true one, β_t^{*Y} is a continuous function of t, while $\hat{\beta}_t^Y$ is a right-continuous function, increasing only by jumps. There is therefore no immediate interpretation of $\hat{\beta}^Y$ as a specific value of the estimated integrated intensity β^{*Y} _inside_ the given model. To interpret $\hat{\beta}^Y$ we therefore need an extension of the model, and this is what in reality will be done in Examples 3.7 and 3.8 below and in Section 4.4. It is quite natural that the extension should include discrete processes with multiple jumps as discussed in Section 2.5, since the structure of $\hat{\beta}^y$ conforms with the accumulated intensities for such processes. The principles for obtaining the extension will be discussed in Section 4.6.

The first important properties of Aalen estimators are given in the next result.

3.3. Theorem. With respect to P_α, the processes $\hat{\beta}^y - \beta^{*y}$, $(y \in E)$, are orthogonal martingales belonging to $M_0^2(P_\alpha)$ with

$$< \hat{\beta}^y - \beta^{*y} >_t = \int_0^t ds\ \alpha^y(s)\ \frac{1}{\tilde{z}_s^y}\ 1_{(\tilde{z}_s^y > 0)}(s)\ .$$

Proof. Write $M^{i,y}$ for the P_α-martingale

$$M_t^{i,y} = N_t^{i,y} - \int_0^t ds\ \alpha^y(s) Z_s^{i,y}\ .$$

By Proposition 1.4, P_α has finite expectations locally, hence $M^{i,y} \in M_0^2(P_\alpha)$ and the $M^{i,y}$ are orthogonal by Theorem 2.2.13. Now observe that

(3.4)
$$\hat{\beta}_t^y - \beta_t^{*y} = \sum_{i=1}^r M_t^{i,y}\left(\frac{1}{\tilde{z}_-^y}\ 1_{(\tilde{z}_-^y > 0)}\right)\ ,$$

where by Theorem 3.2.8, the stochastic integrals $M^{i,y}((\tilde{z}_-^y)^{-1}\ 1_{(\tilde{z}_-^y > 0)})$ are orthogonal martingales in $M_0^2(P_\alpha)$. But then also $\hat{\beta}^y - \beta^{*y}$ belongs to $M_0^2(P_\alpha)$, and by computation, using Theorem 3.2.8 (iii) and Proposition 3.1.19,

$$< \hat{\beta}^y - \beta^{*y} >_t = \sum_{i=1}^r \Lambda_t^{i,y}\left((\tilde{z}_-^y)^{-2}\ 1_{(\tilde{z}_-^y > 0)}\right)$$

$$= \sum_{i=1}^r \int_0^t ds\ \alpha^y(s)\ \frac{z_s^{i,y}}{(\tilde{z}_s^y)^2}\ 1_{(\tilde{z}_s^y > 0)}$$

$$= \int_0^t ds\ \alpha^y(s)\ \frac{1}{\tilde{z}_s^y}\ 1_{(\tilde{z}_s^y > 0)}$$

for all $y \in E$, while for $y \neq z$

$$< \hat{\beta}^Y - \beta^{*Y} , \hat{\beta}^Z - \beta^{*Z} >_t = 0 .$$ ▌

<u>Remark</u>. For all (stochastic) Lebesgue-integrals above, we have used right-continuous integrands, and to save some - signs in the notation, we shall continue to do so. For the stochastic integrals in (3.4), it is of course critically important that the integrands be left-continuous, hence predictable. ▌

3.5. <u>Corollary</u>. For all t,y, $\hat{\beta}^Y_t$ is an unbiased estimator of $P_\alpha \beta^{*Y}_t$.

<u>Proof</u>. By Theorem 3.3, $\hat{\beta}^Y - \beta^{*Y}$ is a P_α-martingale starting at 0. Hence $P_\alpha(\hat{\beta}^Y_t - \beta^{*Y}_t) = P_\alpha(\hat{\beta}^Y_0 - \beta^{*Y}_0) = 0$. ▌

The variation of $\hat{\beta}^Y$ around β^{*Y} may be measured by the mean squared error function

$$\sigma^2_y(t) = P_\alpha(\hat{\beta}^Y_t - \beta^{*Y}_t)^2 \qquad\qquad (t \geq 0),$$

(which is not the variance of $\hat{\beta}^Y_t$) .

3.6. <u>Proposition</u>. For all t,y ,

$$\hat{\sigma}^2_y(t) = \int_{(0,t]} \tilde{N}^Y(ds) \frac{1}{(\tilde{Z}^Y_{s-})^2} 1_{(\tilde{Z}^Y_{s-} > 0)}(s)$$

is an unbiased estimator of $\sigma^2_y(t)$.

<u>Proof</u>. From Theorem 3.3 we have that

$$\sigma^2_y(t) = P_\alpha(\hat{\beta}^Y_t - \beta^{*Y}_t)^2$$

$$= P_\alpha < \hat{\beta}^Y - \beta^{*Y} >_t$$

$$= P_\alpha \int_0^t ds\, \alpha^Y(s) \frac{1}{\tilde{Z}^Y_s} 1_{(\tilde{Z}^Y_s > 0)} .$$

By Theorem 3.2.8 (i),

$$P_\alpha N_t^{i,y} \left((\tilde{Z}_-^y)^{-2} \, 1_{(\tilde{Z}_-^y > 0)} \right) = P_\alpha \int_0^t ds \; \alpha^y(s) \; \frac{Z_s^{i,y}}{(\tilde{Z}_s^y)^2} \, 1_{(\tilde{Z}_s^y > 0)}$$

and summing on i, the conclusion follows immediately. ▌

3.7. <u>Example</u>. In the full one-dimensional Aalen model with

$$\lambda_{t-} = \mu(t-)(r-N_{t-})^+$$

cf. Example 1.6, the Aalen estimator for

$$\beta_t^* = \int_0^t ds \; \mu(s) 1_{(N_s < r)}$$

becomes, writing $R_{s-} = (r-N_{s-})^+$

$$\hat{\beta}_t = \int_{(0,t]} N(ds) \; \frac{1}{R_{s-}} \, 1_{(N_{s-} < r)}$$

$$= \sum_{k=1}^{N_t \wedge r} \frac{1}{r-(k-1)}$$

which is the so-called <u>Nelson estimator</u> of the integrated intensity.
(Notice that $R_{\tau_{k-}} = r-(k-1)$).

Thinking of τ_k as the time of the k'th observed death among r
individuals with i.i.d. lifetimes (with intensity μ), it is seen that
at τ_k, $\hat{\beta}$ has a jump of size $\frac{1}{R_{\tau_k-}}$ where R_{s-} is the number of individuals alive or at risk just before time s.

If one interprets the Nelson estimator as the accumulated intensity for a purely discrete distribution, one finds by (2.5.5) that the corresponding survivor function $\hat{G}$ has atoms at $\tau_1, \cdots, \tau_r$ with

$$\hat{g}(\tau_k) = \frac{1}{r-(k-1)} \prod_{\ell=1}^{k-1} \left(1 - \frac{1}{r-(\ell-1)} \right) = \frac{1}{r},$$

i.e. $\hat{G}$ is the empirical survivor function. For $\hat{G}$ itself we have
the product integral representation

$$\hat{G}(t) = \prod_{s \leq t} \left(1 - \frac{\Delta N_s}{R_{s-}}\right),$$

for which reason $\hat{G}$ is called a _product-limit estimator_. Notice that although the Aalen-Nelson estimator only estimates μ on the interval $[0,\tau_r]$, the estimator $\hat{G}$ is a completely specified survivor function for a probability on $(0,\infty]$, corresponding to the intensity being 0 on $(\tau_r,\infty]$. Below we shall see an example where a survivor function is not estimated everywhere.

The estimator of the mean squared error function of $\hat{\beta}$ is

$$\hat{\sigma}^2(t) = \sum_{k=1}^{N_{t \wedge r}} \frac{1}{(r-(k-1))^2} . \qquad\qquad \mathbf{I}$$

3.8. _Example_. Consider the product Aalen model of r one-dimensional processes having intensities

$$\lambda_{t-}^i = \mu(t-)1_{(\tau_1^i \wedge u_i \geq t)},$$

cf. Examples 1.7 and 2.2. So this is the situation with r i.i.d. lifetimes, where the lifetime of the i'th individual is observed only if it does not exceed a fixed censoring time u_i. Denote by R_{t-} the size of the risk set just before time t, i.e. the number of individuals under observation (non-censored and alive) just before t. Formally

$$R_{t-} = \sum_i 1_{(\tau_1^i \wedge u_i \geq t)} .$$

By Definition 3.1, the Aalen estimator (Nelson estimator) of

$$\beta^*(t) = \int_0^t ds\, \mu(s) 1_{(R_s > 0)}(s)$$

is

$$\hat{\beta}(t) = \int_{(0,t]} \tilde{N}(ds) \frac{1}{R_{s-}} 1_{(R_{s-} > 0)}(s) ,$$

where the interpretation of $\tilde{N}_t$ is the number of deaths actually observed on $[0,t]$.

With $\hat{\beta}$ we get an estimator of the integrated intensity for μ on the interval $[0,\tau]$ where $\tau = \max_i \{\tau_1^i \wedge u_i\}$. As in Example 3.7 we interpret $\hat{\beta}$ as the accumulated intensity for a purely discrete survivor function $\hat{G}$, and use $\hat{G}$ as estimator for the unknown survivor function for the lifetimes of the individuals. The intensity for $\hat{G}$ is

$$\frac{\hat{g}(t)}{\hat{G}(t-)} = \frac{\Delta \tilde{N}_t}{R_{t-}} \ ,$$

so the atoms of $\hat{G}$ are located at the observed (non-censored) times of death, and by (2.5.4)

$$\hat{G}(t) = \prod_{s \leq t} \left(1 - \frac{\Delta \tilde{N}_s}{R_{s-}} \right) .$$

This product-limit estimator is exactly the Kaplan-Meier estimator discussed in Example 1.2.7.

We have defined $\hat{G}$ only on the interval $[0,\tau]$, and as we shall now see, $\hat{G}$ need not be completely specified as the survivor function for a probability on $(0,\infty]$. In fact $\hat{G}(\tau) = 0$ iff τ is one of the τ_1^i , i.e. corresponds to an observed death, because $\hat{G}(\tau) = 0$ iff $\Delta N_\tau = R_{\tau-} > 0$. If τ is one of the u_i , $\hat{G}(\tau) > 0$ and nothing is said about where the remaining probability mass should be placed on $(\tau,\infty]$.

4.4. Estimation in Markov chains.

In Example 2.1.19 we saw that a continuous time Markov chain on a finite state-space with a fixed initial state, may be viewed as a multi-dimensional counting process by counting the number of transitions between pairs of states. And in Example 1.9 it was shown that the intensity of the counting process has a multiplicative structure.

Now consider a product model corresponding to r independent realizations of a Markov chain, each realization starting at some fixed state allowed to differ from chain to chain.

We denote the state-space of the chains by S, write i,j for elements of S and use index ℓ to refer to the ℓ'th of the r chains.

For $i,j \in S$, $\mu_i(t)$ is the intensity function for the waiting time distribution in i and $\pi_{ij}(t)$ is the jump probability from i to j at time t. Finally we put

(4.1)
$$\alpha_{ij}(t) = \begin{cases} - \mu_i(t) & i = j \\ \mu_i(t)\pi_{ij}(t) & i \neq j \end{cases}$$

so for $i \neq j$, $\alpha_{ij}(t)$ is the transition intensity from i to j at time t and for any t, the matrix $(\alpha_{ij}(t))_{i,j \in S}$ is the intensity matrix of the Markov chains at time t, characterized by having all off-diagonal elements ≥ 0 and all row sums $= 0$.

By Example 1.9, the intensity process for the product of the r counting processes associated with the chains becomes $\lambda_- = (\lambda_-^{\ell,ij})$ where

$$\lambda_{t-}^{\ell,ij} = \alpha_{ij}(t-)1_{(J_{t-}^\ell = i)}$$

for $1 \leq \ell \leq r$, $i \neq j \in S$. Here $J_t^\ell = J_{\overline{N}_t^\ell}^\ell$ is the position of the ℓ'th subchain at time t.

In the statistical model each α_{ij} (for $i \neq j$) is allowed to vary freely in A corresponding to $\mu_i \in A$ and the π_{ij} right-continuous with left-limits, non-negative and satisfying $\sum_{j \neq i} \pi_{ij}(t) = 1$.

Now define

$$S_t^i = \sum_{\ell=1}^{r} 1_{(J_t^\ell = i)}$$

so that S_t^i is the number of chains in state i at time t. Then by Definition 3.1, the Aalen estimator of

(4.2) $$\beta^{*ij}(t) = \int_0^t ds\, \alpha_{ij}(s) 1_{(S_s^i > 0)}(s)$$

for $i \neq j$, $t \geq 0$ is

(4.3) $$\hat{\beta}^{ij}(t) = \int_{(0,t]} \tilde{N}^{ij}(ds) \frac{1}{S_{s-}^i} 1_{(S_{s-}^i > 0)}(s)$$

where $\tilde{N}_t^{ij} = \sum_{\ell=1}^{r} N_t^{\ell,ij}$ is the total number of transitions from i to j on $[0,t]$ in all the chains.

The properties of the estimator are given in Theorem 3.3 and will not be discussed further here. Instead we shall in detail discuss the interpretation of the estimator, and show in particular how it may be used to obtain natural estimators of the common transition probabilities for the r chains.

If P is any probability in the product model, the distribution of any of the subchains with respect to P is given by

(4.4) $$P(\tau_{n+1}^\ell > t \mid \tau_1^\ell, \ldots, \tau_n^\ell, J_1^\ell, \ldots, J_n^\ell) = \exp\left(-\int_{\tau_n^\ell}^t ds\, \mu_{J_n^\ell}(s)\right) ,$$

$$P(J_{n+1}^\ell = j \mid \tau_1^\ell, \ldots, \tau_n^\ell, \tau_{n+1}^\ell, J_1^\ell, \ldots, J_n^\ell) = \pi_{J_n^\ell, j}(\tau_{n+1}^\ell) ,$$

cf. Example 2.1.19. Here of course τ_n^ℓ is the time of the n'th jump in the ℓ'th subchain (or sub counting process) and J_n^ℓ is the state reached by the ℓ'th chain at τ_n^ℓ (corresponding to a jump in component (J_{n-1}^ℓ, J_n^ℓ) of the ℓ'th counting process).

Equation (4.4) is naturally only valid for $t \geq \tau_n^\ell$. Notice that since in the model all $\mu_i \in A$, the probability (4.4) is by assumption always > 0 on the set $(\tau_n^{\ell} < \infty)$.

In order to estimate the transition probabilities we shall need an expression for these in terms of the α_{ij}. Writing

$$p_{ij}(s,t) = P(J_t^{\ell} = j | J_s^{\ell} = i)$$

for $i,j \in S$, $s \leq t$ we have that

(4.5)
$$p_{ij}(s,t) = \sum_{n \geq 0} p_{ij}^{(n)}(s,t)$$

where $p_{ij}^{(n)}(s,t)$, the probability of moving from i at time s to j at time t with n jumps on the way, is determined recursively by

(4.6)
$$p_{ij}^{(0)}(s,t) = \delta_{ij} \exp\left(-\int_s^t du\, \mu_i(u)\right),$$

(4.7)
$$p_{ij}^{(n+1)}(s,t) = \sum_{k \neq i} \int_s^t du\, \alpha_{ik}(u) e^{-\int_s^u \mu_i}\, p_{kj}^{(n)}(u,t).$$

We shall now obtain estimators for the transition probabilities, by estimating first the distribution of the waiting times in each state i and the jump probabilities π_{ij}, and then insert these estimators in (discrete) analogues of (4.6), (4.7) and (4.5).

(It is appropriate to point out, that the perhaps most obvious estimator of $p_{ij}(s,t)$,

$$\overset{\vee}{p}_{ij}(s,t) = \frac{1}{S_r^i(s)} \sum_{\ell=1}^r 1_{(J_s^{\ell} = i, J_t^{\ell} = j)},$$

which is the observed frequency of transitions from i to j over $(s,t]$, among the chains in state i at time s, cannot be used. Reason: the matrices $\overset{\vee}{P}(s,t) = (\overset{\vee}{p}_{ij}(s,t))$ do not satisfy the Chapman-Kolmogorov equations, hence do not describe a Markov chain).

Since $\sum_{j \neq i} \alpha_{ij} = \mu_i$, summing on j in (4.2) and (4.3) we obtain

$$\hat{\beta}^{i\cdot}(t) = \int_{(0,t]} \widetilde{N}^{i\cdot}(ds)\, \frac{1}{S_{s-}^i}\, 1_{(S_{s-}^i > 0)}(s)$$

as estimator of the integrated intensity

(4.8)
$$\beta^{*i\cdot}(t) = \int_0^t ds\ \mu_i(s) 1_{(S_s^i > 0)}(s)$$

where $\tilde{N}_t^{i\cdot} = \sum_{j \neq i} \tilde{N}_t^{ij}$ is the total number of transitions from i on $(0,t]$ in all the chains.

Denoting by $G^{(i)}$ the survivor function with intensity μ_i, a natural estimator for $G^{(i)}$ is $\hat{G}^{(i)}$, the purely discrete survivor function with accumulated intensity $\hat{\beta}^{i\cdot}$. Then $\hat{G}^{(i)}$ has atoms at the timepoints where a transition from i in one of the subchains is observed to occur, and for t such a time the obvious estimator of the jump probabilities $\pi_{ij}(t-)$ for all j with $j \neq i$, is the frequency

$$\hat{\pi}_{ij}(t) = \frac{\Delta\tilde{N}_t^{ij}}{\Delta\tilde{N}_t^{i\cdot}}$$

of observed transitions from i to j among all transitions from i. Under the model no two subchains can jump simultaneously so the $\hat{\pi}_{ij}$ take a trivial form: at any t where a transition from i is observed, only one such transition takes place and then $\hat{\pi}_{ij}(t) = 1$ if that transition is from i to j.

Because the $\hat{G}^{(i)}$ are purely discrete with atoms at the observed jump times, when estimating the transition probabilities $p_{ij}(s,t)$ it is immaterial how the $\hat{\pi}_{ij}(t)$ are defined for t not an observed jump time.

We shall now discuss some non-trivial problems arising when understanding the estimator $\hat{G}^{(i)}$.

It is perfectly natural that μ_i (or $G^{(i)}$, as seen by (4.8)), can only be estimated at timepoints t where at least one of the chains is observed to be in state i. But that set of timepoints will in general be a disjoint union of intervals, only finitely many such intervals intersecting $[0,t]$ for any $t > 0$. Suppose that

$[\rho_1,\sigma_1)$, $[\rho_2,\sigma_2),\cdots$ are these intervals where $\sigma_n < \rho_{n+1}$. (Of course the ρ_n and σ_n are random, depending on the observation of the chains. In the sequel one may think of that observation as fixed, the estimation being performed on the basis of the given observation). Interpreting $\hat{\beta}^{i\cdot}$ as the accumulated intensity for $\hat{G}^{(i)}$, it is seen that on the first interval $(\rho_1,\sigma_1]$,

$$\hat{G}^{(i)}(t) = \prod_{\rho_1 < s \leq t}\left(1 - \frac{\Delta\tilde{N}^{i\cdot}_s}{S^i_{s-}}\right) \qquad (\rho_1 \leq t \leq \sigma_1).$$

But $\Delta\tilde{N}^{i\cdot}_{\sigma_1} = S^i_{\sigma_1-} > 0$ because none of the chains are in i at time σ_1, hence $\hat{G}^{(i)}(\sigma_1) = 0$ and $\hat{G}^{(i)}$ is the survivor function for a unique probability on $(\rho_1,\sigma_1]$. Coming to the next interval, we therefore have to estimate $\hat{G}^{(i)}$ afresh by

$$\hat{G}^{(i)}(t) = \prod_{\rho_2 < s \leq t}\left(1 - \frac{\Delta\tilde{N}^{i\cdot}_s}{S^i_{s-}}\right) \qquad (\rho_2 \leq t \leq \sigma_2).$$

Proceeding in this manner we see that $\hat{G}^{(i)}$ is defined on $\bigcup_n [\rho_n,\sigma_n]$ and on each interval $[\rho_n,\sigma_n]$, $\hat{G}^{(i)}$ is the survivor function for a probability on $[\rho_n,\sigma_n]$, in particular $\hat{G}^{(i)}(\sigma_n) = 0$, $\hat{G}^{(i)}(\rho_n) = 1$. Thus the estimator for $G^{(i)}$, which by the model is specified to be a survivor function with $G^{(i)}(t) > 0$ for all $t > 0$, is a function $\hat{G}^{(i)}$ which on each of several disjoint closed intervals agrees with the survivor function for a probability concentrated on that interval.

For the Markov chains discussed in Example 2.1.19 we only had one survivor function associated with each state. But it is still true that the $\hat{G}^{(i)}$ and the $\hat{\pi}_{ij}$ are the conditional jump time distributions and conditional jump probabilities for a unique Markov chain probability $\hat{P}$ on the space of right-continuous, left-limit paths taking values in the state-space S, namely

$$(4.9) \qquad \hat{P}(\tau_{n+1} > t | \tau_1, \cdots, \tau_n, J_1, \cdots, J_n) = \frac{\hat{G}^{(J_n)}(t)}{\hat{G}^{(J_n)}(\tau_n)} \quad ,$$

$$(4.10) \qquad \hat{P}(J_{n+1} = j | \tau_1, \cdots, \tau_{n+1}, J_1, \cdots, J_n) = \hat{\pi}_{J_n j}(\tau_{n+1}) \quad ,$$

cf. (2.1.20) and the identity following it. (One should of course con-
vince oneself that everything fits so that although the $\hat{G}^{(i)}$ and
$\hat{\pi}_{ij}$ are not defined everywhere, $\hat{P}$ is uniquely determined by (4.9)
and (4.10) and the problem of dividing by 0 in (4.9) does not arise.
Also one should check that $\hat{P}$ is Markov!)

It is natural that the estimators of the unknown transition probabilities $p_{ij}(s,t)$ should be the transition probabilities $\hat{p}_{ij}(s,t)$
for the Markov probability $\hat{P}$. But the $\hat{p}_{ij}$ are of course determined
by the $\hat{G}^{(i)}$ and $\hat{\pi}_{ij}$, the expressions being analogous to (4.5),
(4.6) and (4.7):

$$\hat{p}_{ij}(s,t) = \sum_{n \geq 0} \hat{p}_{ij}^{(n)}(s,t)$$

where

$$\hat{p}_{ij}^{(0)}(s,t) = \delta_{ij} \frac{\hat{G}^{(i)}(t)}{\hat{G}^{(i)}(s)}$$

$$\hat{p}_{ij}^{(n+1)}(s,t) = \sum_{k \neq i} \int_{(s,t]} \frac{\hat{G}^{(i)}(du)}{\hat{G}^{(i)}(s)} \hat{\pi}_{ik}(u) \hat{p}_{kj}^{(n)}(u,t) .$$

In order to find $\hat{p}_{ij}$ we shall only use these identities over
small intervals where they give something very simple, and then com-
bine this information with the fact that we know that the transition
matrices $\hat{P}(s,t) = (\hat{p}_{ij}(s,t))_{i,j \in S}$ satisfy the Chapman-Kolmogorov
equations:

$$\hat{P}(s,t) = \hat{P}(s,u)\hat{P}(u,t) \qquad (s \le u \le t).$$

To get expressions for $\hat{P}(s,t)$ valid for all $s \le t$, we now for convenience define $\hat{G}^{(i)}(s) = 1$ for all s where it was not defined previously or was equal to 0. (By the remarks above $\hat{P}(J_s = i) = 0$ for all such s, so the resulting $\hat{P}(s,t)$ will be a set of transition probabilities for $\hat{P}$).

So now fix $s \ge 0$ and let $d > s$ be the first point to the right of s which is an atom for one of the $\hat{G}^{(i)}$. Then for $s \le t < d$, the $\hat{P}$-chain cannot jump on $(s,t]$ and therefore

$$\hat{P}_{ij}(s,t) = \hat{p}_{ij}^{(0)}(s,t) = \delta_{ij} \frac{\hat{G}^{(i)}(t)}{\hat{G}^{(i)}(s)} = \delta_{ij},$$

while if $t = d$ at most one jump can occur so that

$$\hat{P}_{ij}(s,d) = \hat{p}_{ij}^{(0)}(s,d) + \hat{p}_{ij}^{(1)}(s,d)$$

$$= \delta_{ij} \frac{\hat{G}^{(i)}(d)}{\hat{G}^{(i)}(s)} + \frac{\hat{g}^{(i)}(d)}{\hat{G}^{(i)}(s)} \hat{\pi}_{ij}(d)$$

$$= \delta_{ij} + (\hat{\pi}_{ij}(d) - \delta_{ij}) \frac{\hat{g}^{(i)}(d)}{\hat{G}^{(i)}(s)}.$$

But here

$$\frac{\hat{g}^{(i)}(d)}{\hat{G}^{(i)}(s)} = \frac{\hat{g}^{(i)}(d)}{\hat{G}^{(i)}(d-)} = \frac{\Delta \tilde{N}_d^{i\cdot}}{s_{d-}^i}$$

if i is that state for which d is an atom for $\hat{G}^{(i)}$, and then

$$\hat{\pi}_{ij}(d) = \frac{\Delta \tilde{N}_d^{ij}}{\Delta \tilde{N}_d^{i\cdot}}.$$

Introducing therefore the estimated accumulated intensity matrix $\hat{Q}(t) = (\hat{q}_{ij}(t))_{i,j \in S}$ by

$$\hat{q}_{ij}(t) = \sum_{u \le t} \left(\hat{\pi}_{ij}(u) - \delta_{ij} \right) \frac{\hat{g}^{(i)}(u)}{\hat{G}^{(i)}(u-)}$$

$$= \sum_{u \leq t} \left(\frac{\Delta \tilde{N}_u^{ij}}{S_{u-}^i} - \delta_{ij} \frac{\Delta \tilde{N}_u^{i \cdot}}{S_{u-}^i} \right)$$

(where in the last sum the u'th term is 0 if S_{u-}^i and therefore also $\Delta \tilde{N}_u^{i \cdot}$ is 0), it is seen that for $s \leq t \leq d$

$$\hat{P}(s,t) = \prod_{s < u \leq t} (I + \Delta \hat{Q}(u)).$$

Combining this with the Chapman-Kolmogorov equations we are forced to take

$$\hat{P}(s,t) = \prod_{s < u \leq t} (I + \Delta \hat{Q}(u)) \qquad (s \leq t)$$

for arbitrary values of s and t, which is then the desired estimator of the transition probabilities. The estimator has the form of a matrix-valued product integral!

Notice that for every t the matrix $\Delta \hat{Q}(t)$ is an intensity matrix, i.e. the off-diagonal elements are ≥ 0 and the row sums are 0. The accumulated intensity

$$\sum_{u \leq t} \Delta \hat{Q}(u)$$

is the Aalen estimator of the integrated intensity matrix

$$\left(\int_0^t ds \ \alpha_{ij}(s) \right)_{i,j \in S},$$

with the α_{ij} given by (4.1).

One final remark: if the number r of observed chains is large and t > 0 is fixed, one will expect that at least one of the subchains is in any given state i at any time $s \leq t$. In that case of course $\hat{G}^{(i)}$ is estimated on [0,t] by just one survivor function.

4.5. The Cox regression model.

We have seen that a model involving i.i.d. survival times without or with censoring can be treated as a (product) Aalen model (Examples 1.6, 1.7, 2.2, 2.8, 3.7, 3.8). We shall now discuss an extension of these models due to Cox (1972).

The basic idea is to consider independent survival times no longer assumed to be identically distributed, but such that the distribution of any individual depends on a set of covariates characteristic of that individual.

To take a specific situation, suppose one is interested in the survival time for patients suffering from some illness. Measuring the (stochastic) survival time for a patient from the time the illness was diagnostized, its distribution will typically depend on the general constitution of that patient and other factors, which may well be observed and taken into account, e.g. age, sex, past history etc. Also, if the patients are subjected to different treatments (one may wish to compare the effects of the various treatments), the treatment is a covariate that should enter the model.

The model is now defined as follows: suppose there are r individuals with independent lifetimes $X_1, \cdots, X_r$, the distributions of which all have smooth densities. Assume also, that the values of p covariates are recorded for each individual, resulting in a vector $z^i \in \mathbb{R}^p$ for individual i, $1 \leq i \leq r$. (It is convenient to assume the observed values of the covariates to have this form, even though some of them (e.g. the sex of a patient) may be qualitative and not quantitative in nature). Then the intensity function for the distribution of X_i has the form

$$(5.1) \qquad \mu(t)h(\beta, z^i)$$

where $\mu \in A$, β is a vector of unknown real parameters belonging to some domain Θ, and h is a given non-negative, measurable function.

Cox discusses the special case

(5.2) $$h(\beta,z) = e^{(\beta,z)}$$

in particular, where $\beta \in \mathbb{R}^p$ and $(\beta,z) = \sum_j \beta_j z_j$ is the inner product of vectors β and z . For our discussion here, the choice of
h is unimportant.

As defined, the model specifies that the survival intensity for
individual i is proportional to an unknown intensity μ (the <u>baseline
hazard</u>), the proportionality factor depending on a set of unknown parameters and the observed values for i of the covariates. Therefore
the survivor function G^i for the distribution of X_i is a power

(5.3) $$G^i(t) = (G(t))^{h(\beta,z^i)}$$

of the survivor function G with intensity μ .

It should be stressed, that for simplicity we are assuming here
that the values z^i of the covariates are non-stochastic and independent of time. This is the simplest version of the Cox model, of which
other variations may be obtained by allowing the z^i to depend on t ,
as is relevant to many applications (but then one does not have a proportional hazards model as above), and/or be stochastic. Also censoring
may be introduced.

Notice that with the exponential h from (5.2), the baseline
hazard is the intensity for the survivor function of an individual with
$z^i = 0$.

Now consider the r-dimensional counting process $K = (K^1, \cdots, K^r)$
where

$$K_t^i = 1_{(X_i \leq t)} .$$

The canonical counting process generated by K has intensity
$\underline{\lambda} = (\lambda_-^i)_{1 \leq i \leq r}$, where

(5.4) $$\lambda^i_{t-} = \mu(t-)h(\beta,z^i)(1-N^i_{t-})^+ .$$

When μ varies freely in A and β in Θ , this gives a multiplicative intensity model, with the (parametric) h-factor the novelty of this model compared to the standard product Aalen models. From now on assume $h > 0$ (as is the case if h is given by (5.2)). Of course, for β known, we have an ordinary Aalen model, and therefore according to Definition 3.1 estimate the integrated baseline hazard (and this is where the assumption $h > 0$ is useful)

$$\int_0^t ds\ \mu(s)1_{(\tilde{N}_s<r)}(s) ,$$

where $\tilde{N} = \sum\limits_{i=1}^{r} N^i$, by

(5.5) $$\int_{(0,t]} \tilde{N}(ds)\ \frac{1}{\sum\limits_{i\in R(s-)} h(\beta,z^i)}\ 1_{(\tilde{N}_{s-}<r)}(s) ,$$

writing $R(s-) = \{i: N^i_{s-} = 0\}$ for the risk set of individuals alive just before s .

The main problem however, is to investigate the influence of the covariates on the survival distribution, and for this we must estimate β . (Then, inserting this estimator in (5.5), we obtain an estimator for the integrated baseline hazard).

We shall briefly discuss the problem of estimating β , assuming that the process is observed until all individuals have died. Now consider

$$C = \prod_{k=1}^{r} \frac{h(\beta,z^{Y_k})}{\sum\limits_{i\in R_k} h(\beta,z^i)} ,$$

where, with $\tau_1<\cdots<\tau_r$ the times of death, Y_k is the individual dying at τ_k and $R_k = R(\tau_k-)$ is the set of $r-(k-1)$ individuals alive immediately before time τ_k .

The quantity C is Cox's <u>partial likelihood</u>, and Cox's proposal for estimating β now consists in finding the value of β which maximises C .

The label 'partial likelihood' derives from the following fact:
for $1 \le k \le r$, $i \in R_k$ we have

(5.6)
$$P(Y_k = i | F_{\tau_k^-}) = \frac{h(\beta, z^i)}{\sum\limits_{j \in R_k} h(\beta, z^j)} ,$$

namely, in terms of the usual notation the conditional probability is

$$\pi_{k-1} \prime_{\xi_{k-1}} (\tau_k^-, i) = \frac{\lambda^i_{\tau_k^-}}{\sum\limits_j \lambda^j_{\tau_k^-}} ,$$

which using (5.4) immediately gives (5.6).

So C is the product over k of the right-hand side of (5.6),
inserting Y_k for i in the k'th factor. If this product is multipl-
ied by

$$\prod_{k=1}^{r} P(\tau_k \in dt_k | F_{\tau_{k-1}}) = \mu(t_k) \sum_{i \in R_k} h(\beta, z^i) \exp\left(-\int_{t_{k-1}}^{t_k} \mu \sum_{R_k} h\right)$$

we obtain the joint probability density (total likelihood) of
$(\tau_1, \cdots, \tau_r, Y_1, \cdots, Y_r)$:

(5.7)
$$P(\tau_1 \in dt_1, \cdots, \tau_r \in dt_r, Y_1 = i_1, \cdots, Y_r = i_r)$$
$$= \prod_{k=1}^{r} \mu(t_k) h(\beta, z^{i_k}) \exp\left(-\int_{t_{k-1}}^{t_k} \mu \sum_{R_k^0} h\right) dt_k$$

for $t_1 < \cdots < t_r$, $i_1, \cdots, i_r$ a permutation of $1, \cdots, r$ and
$R_k^0 = \{i_k, \cdots, i_r\}$. Hence C is partial in the sense that it contains
only some of the factors that enter into the total likelihood.

Because we have assumed the z^i to be time-independent, the right-
hand side of (5.6) does not depend on $(\tau_1, \cdots, \tau_k)$, hence equals

$$P(Y_k = i | Y_1, \cdots, Y_{k-1}),$$

so that for $i_1, \cdots, i_r$ a permutation of $1, \cdots, r$

$$P(Y_1 = i_1, \cdots, Y_r = i_r) = \prod_{k=1}^{r} \frac{h(\beta, z^{i_k})}{\sum\limits_{j \in R_k^0} h(\beta, z^j)} \quad .$$

Thus C is the ordinary likelihood based on observation of $Y_1, \cdots, Y_r$ alone, and estimating β from C amounts to ignoring all information concerning the times of death, taking into account only the order in which deaths occurred. (But if the z^i depend on time, then C does not have an interpretation as a marginal or conditional likelihood).

Complete observation of the process is equivalent to observation of $\tau_1, \cdots, \tau_r, Y_1, \cdots, Y_r$. As is seen from (5.7), the reduction to $Y_1, \cdots, Y_r$ is not sufficient (not even if μ is known), so some information is lost when estimating β from C alone. The main justification for using C is that asymptotically the estimator for β is good (results to this effect exist, but will not be discussed here), and that it is fairly easy to carry out the estimation by iterations.

When in the next section we discuss maximum-likelihood estimation in Aalen models, we shall also briefly return to the Cox model, and see that maximum-likelihood methods lead to a likelihood for estimating β , which is different from C .

One final remark: for the problem of estimating β , the baseline hazard μ appears as a nuisance parameter. Since presumably most of the information in $(\tau_1, \cdots, \tau_r)$ concerns μ rather than β , one might consider the conditional likelihood of $(Y_1, \cdots, Y_r)$ given $(\tau_1, \cdots, \tau_r)$ and use this for estimating β . However, this conditional likelihood is not simple, and certainly not the same as C .

4.6. Maximum-likelihood estimation in Aalen models.

The Aalen estimators were introduced by definition, and although they were shown to be unbiased in some sense, no argument has been presented to show why they are better than other types of estimators.

To justify the use of Aalen estimators one may for instance show that they have desirable properties asymptotically: this is the theme of the next chapter. But one may also try to show that they are maximum-likelihood estimators, and in this section we shall discuss some of the problems one must solve, if a workable maximum-likelihood theory is to be developed, and also see how the theory works in simple examples.

Consider the product model $P = P(z)$ from Definition 4.2.1 of r counting processes with common type-set E. Observing the product process on $[0,t]$ leads to the model $P^t = P^t(z)$ consisting of the restrictions to F_t of the $P \in P$. As was shown in the beginning of the proof of Theorem 2.3(a), the likelihood function for P^t is proportional to

$$(6.1) \qquad L_t^* = \exp\left(-\sum_y \int_0^t ds\ \alpha^y(s) \tilde{z}_s^y\right) \prod_{y \in E} \prod_{k=1}^{\tilde{N}_t^y} \alpha^y(\tilde{\tau}_k^y-).$$

The first step is now to see what happens when one attempts to maximise L_t^* as a function of the unknown α^y.

But it is easily seen, that L_t^* can be made arbitrarily large, by choosing for instance, for every y, continuous α^y which are 0 everywhere except in tiny intervals around each jump point for the y-components where α^y should be large, decreasing sharply to 0 on either side of the jump point so as to make $\int_0^t \alpha^y$ small.

Thus there is no maximum-likelihood estimator in the original model, but it is important to note that the value ∞ for L_t^* is achie-
ved as the limit of values taken for a sequence of peaked α^y as just
described, where the peak values increase to ∞ and the intervals
where the α^y are strictly positive shrink to the set of jump points
for the observed counting process. One possible interpretation of this
is, that the sequence of CCP(rE)'s corresponding to the considered
sequence of α^y, in some (sofar unspecified) sense converges to a
limit which is something like a discrete CCP(rE) (because it will only
generate jumps, where jumps were actually observed), and which limit
could then be called the maximum-likelihood estimator of the unknown
$P^t \in P$.

We shall make this more precise: we want simply to extend the original model P^t by adjoining some discrete processes, to get a new
model $\overline{P}^t$ in which the maximum-likelihood estimator $\hat{P}^t$ for $\overline{P}^t$
$\in \overline{P}^t$ exists, and then use $\hat{P}^t$ as estimator, also in the original model.

Apart from the objection one may have to estimating a probability
in a given model by a probability not belonging to the model, there are
at least three important problems which must be solved before this
approach can be used. Firstly, because the extension must include discrete processes, we no longer have a dominated family of probability
measures, and can therefore not use standard maximum-likelihood theory.
Secondly, there is the problem of determining $\overline{P}^t$. One possibility
is to consider the path-space on which the probabilities in P^t are
defined (i.e. counting process paths on $[0,t)$), then extend this space
to a new path-space $\overline{W}^t$ by including paths with multiple jumps (cf.
the discussion of the third problem below), then equip $\overline{W}^t$ with a topology and the set of probabilities on $\overline{W}^t$ with the matching weak topology, and then finally define $\overline{P}^t$ as the weak closure of P^t. Thirdly,
as will be apparent from the examples below, the extension must comprise

discrete processes where several components can jump simultaneously and/or each component can have jumps of size > 1 . (Such processes were discussed at the end of Section 2.5). To include such processes will, at the very least, cause technical complications.

The first problem is solved by using the theory of maximum-likelihood estimation for non-dominated models, which implies the following: suppose that the extended model $\overline{P}^t$ has the property that to every $w^t \in \overline{W}^t$ there is a $\overline{P}^t \in \overline{P}^t$ with $\overline{P}^t\{w^t\} > 0$. Then, given the observation w^t , the value $\hat{P}_{w^t}$ of the maximum-likelihood estimator is the (any) probability in $\overline{P}^t$ such that

$$\hat{P}_{w^t}\{w^t\} = \sup_{\overline{P}^t \in \overline{P}^t} \overline{P}^t\{w^t\},$$

i.e. the maximum-likelihood estimator maximises the probability of the observation. In the examples below we shall be in this situation.

As for the second problem and the choice of topology on $\overline{W}^t$, consider the simple example where $r = 1$, $|E| = 1$, so that $\overline{W}^t$ includes (one-dimensional) paths with jumps ≥ 1 . The natural topology on $\overline{W}^t$ (fitting with the examples below) is obtained by viewing each $w^t \in \overline{W}^t$ as an integer-valued, discrete measure on $[0,t]$ and then giving $\overline{W}^t$ the weak topology of these measures. (This topology is coarser than the Skorokhod $D[0,t]$-topology (restricted to $\overline{W}^t$): if w_n is the path in $\overline{W}^t$ that jumps one at $\frac{t}{2} - \frac{1}{n}$ and $\frac{t}{2} + \frac{1}{n}$ and nowhere else, and w is the path jumping two at $\frac{t}{2}$ and nowhere else, then $w_n \to w$ in the weak but not in the Skorokhod topology. For details about the Skorokhod topology, see Appendix 2).

As already stated, the third problem is mainly a technical one. This is due to the observation made in Section 2.5, that the intensity characterizing a discrete counting process with multiple jumps, has a much more complicated structure than the intensity of an ordinary CCPE. We shall say nothing general about the technique of finding the actual extensions.

6.2. <u>Example</u>. Consider r i.i.d. lifetimes as in Examples 1.6 and 2.2. The product model has intensity

(6.3)
$$\lambda^i_{t-} = \mu(t-)(1-N^i_{t-})^+$$

for $1 \le i \le r$, where $\mu \in A$.

The natural extension $\bar{P}$ of P is obtained by still considering i.i.d. lifetimes, only allowing the common distribution of these to be purely discrete. Then several deaths can occur simultaneously, and the extended path-space $\bar{W}$ therefore consists of paths $w = (w^1, \cdots, w^r)$, where several components can jump at the same time.

If μ now denotes the discrete intensity function for the purely discrete survival distribution, the intensity for a discrete probability P on $\bar{W}$ is given by $\lambda = (\lambda^A)_{\emptyset \neq A \subset \{1, \cdots, r\}}$ where

$$\lambda^A_t = P(\Delta N^i_t = 1, \ i \in A , \ \Delta N^i_t = 0 , \ i \notin A | F_{t-})$$

is the component of the intensity corresponding to the individuals in A dying simultaneously. With R_{t-} the risk set at time t , we find that

(6.4)
$$\lambda^A_t = \mu(t)^{|A|}(1-\mu(t))^{|R_{t-}| - |A|} 1_{(A \subset R_{t-})} .$$

(Here $|B|$ is the cardinality of B). It is worthwhile observing, that the intensity for individual i dying is

$$\lambda^i_t = \sum_{A, i \in A} \lambda^A_t = \mu(t) 1_{(i \in R_{t-})} .$$

Thus, since $1_{(i \in R_{t-})} = (1-N^i_{t-})^+,$ the structure (6.3) of λ^i_t is retained by the switch from $\mu \in A$ to a discrete intensity μ .

Suppose now that the process of r lifetimes is followed on the interval $[0,t]$. The observation consists in a random number $\tau_1 < \cdots < \tau_{N^*_t} \le t$ of times at which deaths occurred, and random subsets A_k of $\{1, \cdots, r\}$, A_k denoting the set of individuals dying at τ_k so that $A_k \neq \emptyset$ and the A_k are mutually disjoint. According

to Proposition 2.5.12, the likelihood evaluated at an arbitrary $P \in \bar{P}$
is 0 if $P \in P$ and

$$(6.5) \qquad \left(\prod_{\substack{0 < s \leq t \\ s \neq \tau_1, \cdots, \tau_{N^*_t}}} (1-\mu(s))^{|R(s-)|} \right) \prod_{k=1}^{N^*_t} \mu(\tau_k)^{|A_k|} (1-\mu(\tau_k))^{|R_k| - |A_k|}$$

if $P \in \bar{P} \setminus P$, where we have written $R_k = R_{\tau_k -}$, and used (6.4) and
the fact that

$$\sum_A \lambda^A_t = 1 - (1-\mu(t))^{|R_{t-}|} .$$

So the maximum-likelihood estimator is the probability $\hat{P} \in \bar{P} \setminus P$ corresponding to the discrete intensity $\hat{\mu}$ which maximises (6.5). The
first product is maximised taking $\hat{\mu}(s) = 0$ for $s \neq \tau_1, \cdots, \tau_{N^*_t}$, and
the k'th factor in the second product is maximised taking

$$(6.6) \qquad \hat{\mu}(\tau_k) = \frac{|A_k|}{|R_k|} .$$

If in particular the data is generated by some $P \in P$, then a.s.
$|A_k| = 1$ and

$$\hat{\mu}(\tau_k) = \frac{1}{r-(k-1)}$$

and comparing this with Example 3.7, we see that the accumulated intensity $\sum_{u \leq s} \hat{\mu}(u)$ is exactly the Aalen estimator. So in this simple example, the Aalen estimator is also the maximum-likelihood estimator.
Furthermore, as we saw in Example 3.7, $\hat{\mu}$ is the (discrete) intensity
for the empirical survivor function. By the basic theorem that the maximum-likelihood estimator of a function of the 'parameter' is that
function of the maximum-likelihood estimator, we conclude that in the
extended model, the empirical survivor function is the maximum-likelihood estimator of the unknown survivor function for the common distribution of the r lifetimes. ∎

6.7. _Example_. In continuation of the previous example, suppose we have i.i.d. lifetimes with fixed censoring times $u_1, \cdots, u_r$ as in Examples 1,7, 2.2 and 3.8. The intensity for the continuous model is $\lambda_- = (\lambda^i_-)$ given as in Example 3.8. The discrete processes in the extension are obtained by considering i.i.d. lifetimes which follow a discrete distribution, and retaining the censoring times $u_1, \cdots, u_r$. The discrete intensity λ^A_t for the set A of individuals dying simultaneously still has the form (6.4), only now the risk set R_{t-} consists of the individuals under observation just before t, so that $i \in R_{t-}$ iff $\tau^i_1 \geq t$, $u_i \geq t$.

If the process is followed on $[0,t]$ and deaths are observed at $\tau_1 < \cdots < \tau_{N^*_t}$ with the set A_k of individuals observed to die at τ_k, the likelihood is given by (6.5) and the maximum-likelihood estimator for μ is as in (6.6), which is the Aalen estimator. Comparing with Example 3.8 it follows that the maximum-likelihood estimator for the survivor function of the lifetime distribution, is the Kaplan-Meier estimator. ▌

6.8. _Example_. We shall now use the same kind of approach as in the previous two examples, to discuss a discrete extension of the Cox model treated in Section 4.5.

For the original model, the intensity $\lambda_- = (\lambda^i)$ is given by (5.4). To get the discrete probabilities in the extension, think of a sequence of continuous survivor functions, corresponding to a sequence of baseline hazards, converging weakly to a discrete survivor function G. Then the sequence of survivor functions for individual i converges by (5.3) weakly to $G^{h(\beta, z^i)}$, so that the original power structure of the Cox model is preserved by weak convergence.

It is therefore natural to propose that the discrete P in the extended model should correspond to r independent lifetimes with survivor functions $G^i = G^{h(\beta, z^i)}$, where G is an arbitrary purely discrete survivor function and $\beta \in \Theta$.

With this definition of the extension, consider now a given $P \in \overline{P} \setminus P$ specified by a discrete G with discrete intensity μ , and a given β . The discrete intensity μ^i for the survivor function G^i is

$$\mu^i(t) = \begin{cases} \dfrac{\Delta G^i(t)}{G^i(t-)} & \text{if} \quad \mu(t) > 0 \\ \\ 0 & \text{if} \quad \mu(t) = 0 , \end{cases}$$

and here

$$\frac{\Delta G^i(t)}{G^i(t-)} = 1 - \frac{G^i(t)}{G^i(t-)} = 1 - (1-\mu(t))^{h(\beta,z^i)}$$

since

$$G^i(t) = (G(t))^{h(\beta,z^i)} = \prod_{s \leq t} (1-\mu(t))^{h(\beta,z^i)} .$$

Therefore the intensity process λ^A for the individuals in the set A to die simultaneously becomes

$$\lambda^A_t = 1_{(A \subset R_{t-})} \prod_{i \in A} \mu^i(t) \prod_{i \in R_{t-} \setminus A} (1-\mu^i(t))$$

$$= 1_{(A \subset R_{t-})} \prod_{i \in A} (1-(1-\mu(t))^{h(\beta,z^i)}) \prod_{i \in R_{t-} \setminus A} (1-\mu(t))^{h(\beta,z^i)} .$$

From this one finds that

(6.9) $$\sum_A \lambda^A_t = 1 - \prod_{i \in R_{t-}} (1-\mu(t))^{h(\beta,z^i)} ,$$

while the intensity for individual i dying is

$$\lambda^i_t = \sum_{A, i \in A} \lambda^A_t = 1_{(i \in R_{t-})} \mu^i(t) .$$

Suppose the process is followed on $[0,t]$ with N^*_t deaths observed at $\tau_1 < \cdots < \tau_{N^*_t} \leq t$, the individuals A_k dying at τ_k , where the A_k are mutually disjoint. The likelihood (probability with respect to P of obtaining the observation) is by Proposition 2.5.12, 0 if $P \in P$ and, using (6.9)

$$L_t = \left(\prod_{\substack{s \le t \\ s \ne \tau_1,\ldots,\tau_{N_t^*}}} \prod_{i \in R_{s-}} (1-\mu(s))^{h(\beta,z^i)} \right)$$

$$\times \prod_{k=1}^{N_t^*} \left[\prod_{i \in A_k} (1-(1-\mu(\tau_k))^{h(\beta,z^i)}) \prod_{i \in R_k \smallsetminus A_k} (1-\mu(\tau_k))^{h(\beta,z^i)} \right]$$

if $P \in \overline{P} \smallsetminus P$, writing $R_k = R_{\tau_k-}$. Maximising this in μ and β gives the maximum-likelihood estimators $\hat{\mu}$ and $\hat{\beta}$.

Denote by $\hat{\mu}_\beta$ the value of μ maximising L_t for β fixed. Obviously $\hat{\mu}_\beta(s) = 0$ for $s \ne \tau_1,\ldots,\tau_{N_t^*}$, but apart from that it is not in general possible to find $\hat{\mu}_\beta$ explicitly. If however we suppose that all $|A_k| = 1$, as will be the case if the probability generating the data belongs to the original model P, the part of L_t involving $\mu_k = \mu(\tau_k)$ becomes

$$\left(1 - (1-\mu_k)^{h(Y_k)} \right) (1-\mu_k)^{\Sigma_{k+1}}$$

writing Y_k for the individual dying at τ_k, $h(i) = h(\beta,z^i)$ and

$$\Sigma_{k+1} = \sum_{i \in R_k \smallsetminus A_k} h(i) = \sum_{i \in R_{k+1}} h(i).$$

Elementary calculations now show that

(6.10)
$$1-\hat{\mu}_\beta(\tau_k) = \left(1 - \frac{h(Y_k)}{\Sigma_k} \right)^{\frac{1}{h(Y_k)}},$$

which is not the Aalen estimator.

Next, inserting $\hat{\mu}_\beta$ in L_t gives

$$\hat{L}_t = \sup_\mu L_t = \prod_{k=1}^{N_t^*} \frac{h(Y_k)}{\Sigma_k} \left(1 - \frac{h(Y_k)}{\Sigma_k} \right)^{\frac{\Sigma_{k+1}}{h(Y_k)}}.$$

To find $\hat{\beta}$ one must now maximise this as a function of β. Notice that $\hat{L}_t$ is Cox's partial likelihood (for observation on $[0,t]$)

$$C_t = \prod_{k=1}^{N_t^*} \frac{h(Y_k)}{\Sigma_k}$$

times an extra factor. Thus estimating β from Cox's partial likelihood

does not give the maximum-likelihood estimator. However, it is plausible
and supported by actual numerical examples that under suitable conditions on h and the z^i , the dependence upon β of the extra factor
is negligeable asymptotically, so that for r large the Cox estimator
and the maximum-likelihood estimator are close to each other.

The estimator (6.10) has been proposed by Bailey (1979).

I

It is the exception rather than the rule, that the Aalen estimator and the maximum-likelihood estimator agree. In case the estimators differ, it is of interest to compare their asymptotic behavior
as r, the number of subprocesses in the model, tends to ∞. We
conjecture that the two estimators are asymptotically equivalent.
Since the Aalen estimator is much the simpler to calculate and manage,
it may be preferred to the maximum-likelihood estimator. But the derivation of the maximum-likelihood estimator yields an immediate interpretation of the estimator, where the interpretation of the Aalen
estimator is more a matter of intuition and definition.

Notes.

The full Aalen models of Section 4.1 appear as special cases of the models originally proposed by Aalen (1975), (1978). He does not however, explicitly deal with the product models in Section 4.2.

The definition of the estimators is Aalen's original definition.

A regression model generalizing the multiplicative intensity models presented here, has been proposed by Aalen (1980). There the intensity λ_- takes the form

$$\lambda^y_{t-} = \sum_{j=1}^{q} \alpha^j(t-) Z^{yj}_{t-} \, ,$$

where the Z^{yj} are observed processes, and the α^j are to be estimated.

The Nelson estimator from Examples 4.3.7, 4.3.8 was first proposed by Nelson (1969).

The problem of estimating the transition probabilities for a Markov chain was solved by Aalen and Johansen (1978), Fleming (1978a) and (1978b). For the structure of Markov chains with possibly fixed times of discontinuities, see Jacobsen (1972).

For a detailed discussion of the (topological) extension of Aalen models indicated in Section 4.6, see Jacobsen (1982). An alternative extension has been proposed by Johansen (1981b), and formally this gives the Aalen estimators as maximum-likelihood estimators. In my opinion however, Johansen's extension is unsatisfactory for two reasons: firstly the probabilities in the extension do not describe the same phenomenon as the probabilities in the original model, e.g. a model for one lifetime corresponding to a counting process having one jump of size 1, is replaced by a process having one jump of arbitrarily large size; and secondly the extension is not defined on the part of the sample space corresponding to multiple jumps. This makes impossible any definite interpretation of the estimator.

The estimator (4.6.10) for the discrete intensity in the Cox model, and the resulting $\hat{L}_t$ for estimating β, appears first in Bailey (1979).

The Aalen estimators are examples of martingale estimators or m-estimators. See Rebolledo (1978).

Exercises.

1. Consider the product Aalen model from Definition 4.2.1, and let
   for each y, U^y be a real-valued, predictable and locally uniformly bounded process.

   Define for $t \geq 0$, $y \in E$

   $$\beta_t^{*y}(U^y) = \int_0^t ds \; \alpha^y(s) U_s^y \; 1_{(\widetilde{Z}_s^y > 0)} \quad ,$$

   $$\hat{\beta}_t^y(U^y) = \int_{(0,t]} \widetilde{N}^y(ds) \; \frac{U_s^y}{\widetilde{Z}_{s-}^y} \; 1_{(\widetilde{Z}_{s-}^y > 0)} \quad .$$

   Show that with respect to P_α, the processes $\hat{\beta}^y(U^y) - \beta^{*y}(U^y)$
   are orthogonal martingales with

   $$< \hat{\beta}^y(U^y) - \beta^{*y}(U^y) >_t = \int_0^t ds \; \alpha^y(s) \; \frac{U_s^{y2}}{\widetilde{Z}_s^y} \; 1_{(\widetilde{Z}_s^y > 0)} \quad .$$

   Show that $\hat{\beta}_t^y(U^y)$ is an unbiased estimator of $P_\alpha \beta_t^{*y}(U^y)$, and
   find an unbiased estimator of the mean-squared error

   $$P_\alpha \left(\hat{\beta}_t^y(U^y) - \beta_t^{*y}(U^y) \right)^2 \quad .$$

   If U^y is a given non-random function $u^y: [0,\infty) \to \mathbb{R}$, $\hat{\beta}_t^y(U^y)$
   may be considered an estimator of the integral of $\alpha^y(s) \; u^y(s)$
   over those $s \leq t$ at which information about α^y is available.
   Thus it is quite possible to estimate other functionals of the
   α^y than their integrals. ∎

2. In the text, when discussing a model for censored survival data,
   we have only considered right-censoring, i.e. the value of the
   lifetime is observed only if it does not exceed the value of the
   censoring time.

This exercise shows that for i.i.d. observations observed partially, the censoring must be of this type to give a multiplicative
intensity model.

Consider r i.i.d. lifetimes $X_1, \cdots, X_r$ with unknown intensity
μ , but suppose that the time of death of individual i is observed only if it occurs inside a given Borel subset B_i of
$(0, \infty)$. (So for the model in Example 4.3.8, $B_i = (0, u_i]$). In
counting process terminology, this amounts to observing $K = (K^i)$
where

$$K_t^i = 1_{(X_i \leq t, X_i \in B_i)} .$$

To show that this does not in general give a product Aalen model,
consider <u>left-censoring</u> with $B_i = (u_i, \infty)$, and show that the intensity for the canonical process generated by K^i is

$$\lambda_{t-}^i = \mu(t-) \frac{G_\mu(t)}{G_\mu(t) + F_\mu(u_i)} 1_{(u_i < t \leq \tau_1^i)} ,$$

where

$$G_\mu(t) = 1 - F_\mu(t) = \exp\left(-\int_0^t \mu\right) \qquad \blacksquare$$

3.   Consider the product Aalen model with

$$\lambda_{t-}^i = \mu(t-) 1_{(u_i < t \leq \tau_1^i)} ,$$

for $i = 1, \ldots, r$ with $u_1, \ldots, u_r$ given constants.

Find the distribution of τ_1^i and compare the result with the
model discussed at the end of the previous exercise. $\blacksquare$

5. ASYMPTOTIC THEORY.

5.1. A limit theorem for martingales.

The main purpose of this chapter is to derive asymptotic distribution results for Aalen and product-limit estimators. The proofs will be based on the limit theorem to be quoted in this section.

Consider a sequence $((\Omega_n, A_n, A_{n,t}, \mathbb{P}_n))_{n \geq 1}$, $(\Omega, A, A_t, \mathbb{P})$ of filtered probability spaces, and let (M_n), M be a sequence of locally square-integrable multidimensional martingales with type-set E, M_n defined on Ω_n, M defined on Ω. To be specific this means that $M_n = (M_n^y)_{y \in E}$ is an adapted, right-continuous, left-limit process on $(\Omega_n, A_n, A_{n,t}, \mathbb{P}_n)$ taking values in $\mathbb{R}^E$ with $\mathbb{P}_n M_{n,t}^{y\,2} < \infty$ for all $y \in E$, $t \geq 0$ such that

$$\mathbb{P}_n (M_{n,t}^y | A_{n,s}) = M_{n,s}^y \qquad (y \in E, s \leq t),$$

and similar conditions on M. Suppose also that the components of each M_n, M are orthogonal, i.e. the product $M_n^y M_n^z$ or $M^y M^z$ is a martingale whenever $y \neq z$. Finally, denote by $<M_n^y>$ the unique predictable, increasing, right-continuous, left-limit process such that $M_n^{y\,2} - <M_n^y>$ is a martingale.

(We are here dealing with abstractly defined martingales, and for these we have not discussed concepts such as predictability, orthogonality etc. However, in all the applications we shall make, the spaces $(\Omega_n, A_n, A_{n,t})$ will be the canonical path-space (W^E, F, F_t), so that the definitions and the process theory developed in Chapter 3 is sufficient).

The assertion of the limit theorem is formulated in terms of convergence in distribution of M_n to M, written $M_n \overset{D}{\to} M$. This means that we consider M_n, M as $D[0, \infty)^E$- valued random variables, where $D[0, \infty)^E$ is the (product) space of paths $w = (w^y)_{y \in E} : [0, \infty) \to \mathbb{R}^E$ which are everywhere right-continuous with left-limits, equipped with

the product Skorokhod topology, and then demand that $\mathbb{P}_n f(M_n) \to \mathbb{P} f(M)$
for all bounded, continuous $f: D[0,\infty)^E \to \mathbb{R}$. (For a discussion of the
definition of the Skorokhod topology on the (one-dimensional) function
space $D[0,\infty)$, and the resulting weak convergence of probabilities on
$D[0,\infty)$ and convergence in distribution of right-continuous, left-limit
processes, see Appendix 2).

To identify the limit process, we need the definition below. For
the formulation, let $\Phi = (\Phi^y)_{y \in E}$ be a family of non-decreasing, continuous functions $\Phi^y: [0,\infty) \to [0,\infty)$ with $\Phi^y(0) = 0$.

1.1. <u>Definition</u>. A $\mathbb{R}^E$-valued stochastic process $M = (M_t^y)_{y \in E, t \geq 0}$
defined on a filtered probability space $(\Omega, A, A_t, \mathbb{P})$, is a <u>Gauss-$\Phi$ process with independent increments</u> provided M is adapted with continuous
paths and satisfies

(i)   the components M^y are stochastically independent;

(ii) $(M_t^y - M_s^y)_{y \in E}$ is independent of A_s for $s \leq t$;

(iii) M_t^y is normally distributed with mean 0 and variance $\Phi^y(t)$

        for $y \in E, t \geq 0$. |

In the sequel we shall mostly call such a process simply a Gauss-Φ
process, and omit the bit about independent increments.

It follows from the definition that $\mathbb{P}(M_0^y = 0) = 1$ and that for
$y \in E, s \leq t$ the increment $M_t^y - M_s^y$ is normally distributed
$(0, \Phi^y(t) - \Phi^y(s))$. Also M is a multidimensional, locally square integrable martingale with orthogonal components, and by direct computation
it is found that $M^{y^2} - \Phi^y$ is a martingale so that

(1.2) $$< M^y >_t = \Phi^y(t).$$

The basic example of such a process is the Brownian motion B
with type-set E, corresponding to the case $\Phi^y(t) = t$ for $y \in E$,
$t \geq 0$. It is well-known that B can be realized with continuous paths,

simply as $|E|$ independent copies of the standard one-dimensional Brownian motion. From the existence of B it is easy to argue the existence of the general Gauss-Φ process: just define $M = (M^y)$ by

$$M_t^y = B_{\Phi^y(t)}^y \quad .$$

With Definition 1.1 and the notation from the beginning of the section in mind, we can now state the limit theorem.

1.3. <u>Theorem</u>. Let $(M_n)_{n \geq 1}$ be a sequence of locally square-integrable, multidimensional martingales with type-set E, such that for every n the component martingales M_n^y are orthogonal. Suppose that

(a) $\quad \mathbb{P}_n \underset{s \leq t}{\Sigma} (\Delta M_n^y(s))^2 \, 1_{(|\Delta M_n^y(s)| > \varepsilon)} \underset{n \to \infty}{\to} 0 \quad (y \in E, t \geq 0, \varepsilon > 0),$

and suppose also that there exists a family $\Phi = (\Phi^y)_{y \in E}$ of non-decreasing, continuous functions $\Phi^y : [0, \infty) \to [0, \infty)$ with $\Phi^y(0) = 0$ such that

(b) $\quad \mathbb{P}_n \left(| < M_n^y >_t - \Phi^y(t) | > \varepsilon \right) \underset{n \to \infty}{\to} 0 \quad (y \in E, t \geq 0, \varepsilon > 0).$

Then $M_n \overset{\mathcal{D}}{\to} M$ where M is a Gauss-Φ process with independent increments. ∎

We shall not prove this result here. The proof consists in showing that the finite-dimensional distributions of M_n converge weakly to those of M, and that the sequence of distributions of the M_n is tight, cf. the discussion of weak convergence in Appendix 2. The theorem is due to Rebolledo (1978).

Condition (a) in the theorem is a kind of Lindeberg condition requiring that big jumps for the M_n^y be asymptotically rare. Because of (1.2), condition (b) states that $<M_n^y>$ must at each time instant converge in probability to $<M^y>$.

We shall first illustrate the usefulness of the theorem by three elementary examples.

1.4. <u>Example</u>. Let $(\Omega_n, A_n, A_{n,t}, \mathbb{P}_n) = (W, F, F_t, \Pi_{\mu_n})$ be the path-space for one-dimensional counting processes with Π_{μ_n} the Poisson process with (constant) intensity μ_n (Example 1.2.4). Then, for every n,

$$M_n(t) = \frac{1}{\sqrt{\mu_n}} (N_t - \mu_n t)$$

is a Π_{μ_n}-martingale with

$$\langle M_n \rangle_t = \frac{1}{\mu_n} \langle N - \mu_n t \rangle_t = t,$$

so that condition (b) of Theorem 1.3 is trivially satisfied with $\Phi(t) = t$. If it is further assumed that $\mu_n \to \infty$, then (a) is also met because

$$(\Delta M_n(s))^2 \, 1_{(|\Delta M_n(s)| > \varepsilon)} = 0$$

whenever $\frac{1}{\sqrt{\mu_n}} \le \varepsilon$, all jumps for M_n having size $\frac{1}{\sqrt{\mu_n}}$. Therefore, by Theorem 1.3, if $\mu_n \to \infty$, then M_n converges in distribution to Brownian motion.   ▌

1.5. <u>Example</u>. Let $(\Omega_n, A_n, A_{n,t}, \mathbb{P}_n) = (W, F, F_{nt}, \Pi_\mu)$ for all n with Π_μ the Poisson process with intensity μ. (So the only thing that changes with n is the filtration, and this is just to make M_n adapted). Define

$$M_n(t) = \frac{1}{\sqrt{n\mu}} (N_{nt} - n\mu t).$$

Then M_n is a martingale with $\langle M_n \rangle_t = t$. Since

$$(\Delta M_n(s))^2 \, 1_{(|\Delta M_n(s)| > \varepsilon)} = 0$$

for $\frac{1}{\sqrt{n\mu}} \le \varepsilon$, the conditions of Theorem 1.3 are satisfied, and M_n converges in distribution to Brownian motion.   ▌

Comparing Examples 1.4 and 1.5, we see that two types of asymptotics are feasible: either one considers processes with intensities growing large, or else one observes the same process over large intervals of time. The first type occurs for instance by having a large number of Poisson processes, and then considering their sum. This is the type of asymptotics we shall be considering in the sequel.

1.6. <u>Example</u>. Let $(\Omega_n, A_n, A_{n,t}, \mathbb{P}_n) = (W, F, F_t, P_n)$ where P_n is the inhomogeneous Poisson process with intensity function $\mu_n(t)$. Suppose $t_0 > 0$ given with $\int_0^{t_0} \mu_n \to \infty$ and define

$$M_n(t) = \frac{1}{\sqrt{\int_0^{t_0} \mu_n}} \left(N_t - \int_0^t \mu_n \right) .$$

Then M_n is a martingale with

$$<M_n>_t = \frac{\int_0^t \mu_n}{\int_0^{t_0} \mu_n} .$$

Since condition (a) of Theorem 1.3 is satisfied because $\int_0^{t_0} \mu_n \to \infty$, the theorem implies that if

$$\frac{\int_0^t \mu_n}{\int_0^{t_0} \mu_n} \to \Phi(t)$$

where Φ is continuous and finite, then M_n converges in distribution to the Gauss-Φ process. ▌

5.2. Asymptotic distributions of Aalen estimators.

We shall consider the product Aalen model $P(Z)$ from Definition 4.2.1 of r processes with type-set E. Since we are going to let r vary, we shall write $P_{\alpha,r}$ for a typical member of $P_r = P(Z)$, and amend the notation for integrated intensities and their estimators accordingly. Also, the notation used for processes depending on α will include α, so that e.g. $\beta^{*y}_{\alpha,r}$ will be the notation for the integrated intensity β^{*y} from Section 4.3.

By Definition 4.3.1, the Aalen estimator of

$$\beta^{*y}_{\alpha,r}(t) = \int_0^t ds \; \alpha^y(s) 1_{(\tilde{Z}^y_r(s) > 0)}$$

is

$$\hat{\beta}^y_r(t) = \int_{(0,t]} \tilde{N}^y_r(ds) \; \frac{1}{\tilde{Z}^y_r(s-)} 1_{(\tilde{Z}^y_r(s-) > 0)} .$$

By Theorem 4.3.3, with respect to $P_{\alpha,r}$ the process $M_{\alpha,r} = (M^y_{\alpha,r})_{y\in E}$ where $M^y_{\alpha,r} = \hat{\beta}^y_r - \beta^{*y}_{\alpha,r}$ is a multidimensional martingale with type-set E, the components being orthogonal with

$$(2.1) \qquad < M^y_{\alpha,r} >_t = \int_0^t ds \; \alpha^y(s) \frac{1}{\tilde{Z}^y_r(s)} 1_{(\tilde{Z}^y_r(s) > 0)} .$$

We shall now discuss the asymptotic behavior of $\hat{\beta}_r$ as $r \to \infty$, using Theorem 1.3 as the main tool. Although we shall almost exclusively rely on this convergence theorem, one should have in mind, that standard central limit theorems for sums of independent random variables may be useful, cf. Example 3.11 below!

2.2. __Theorem.__ Suppose there exists a family $\Phi_\alpha = (\Phi^y_\alpha)_{y\in E}$ of non-decreasing, continuous functions $\Phi^y_\alpha : [0,\infty) \to [0,\infty)$ with $\Phi^y_\alpha(0) = 0$, and sequences $(a^y_r)_{r\geq 1}$ of positive numbers with $\lim_{r\to\infty} a^y_r = \infty$ for every $y \in E$, such that for all $(\alpha^y)_{y\in E} \in A^E$

(a) $\quad \lim_{r\to\infty} P_{\alpha,r} \int_0^t ds\ \alpha^y(s)\ \dfrac{a_r^{y\,2}}{\tilde{Z}_r^y(s)}\ 1_{(0<\tilde{Z}_r^y(s)<\frac{a_r^y}{\varepsilon})} = 0 \qquad (y\in E, t\ge 0, \varepsilon>0),$

(b) $\quad \lim_{r\to\infty} P_{\alpha,r}\left(\left|\int_0^t ds\ \alpha^y(s)\ \dfrac{a_r^{y\,2}}{\tilde{Z}_r^y(s)}\ 1_{(\tilde{Z}_r^y(s)>0)} - \Phi_\alpha^y(t)\right| > \varepsilon\right) = 0 \quad (y\in E, t\ge 0, \varepsilon>0).$

Then $(a_r^y(\hat{\beta}_r^y - \beta^{*y}_{\alpha,r}))_{y\in E}$ converges in $P_{\alpha,r}$-distribution as $r\to\infty$ to the Gauss-Φ_α process with independent increments.

<u>Proof.</u> We shall apply Theorem 1.3 to the sequence $(a_r^y\ M^y_{\alpha,r})$ of martingales. Since

$$< a_r^y\ M^y_{\alpha,r}> = a_r^{y\,2} < M^y_{\alpha,r}>,$$

it is clear from (2.1), that (b) is a direct translation of condition (b) from Theorem 1.3. The quantity from condition (a) there becomes

$$P_{\alpha,r}\ \sum_{s\le t} a_r^{y\,2}(\Delta M^y_{\alpha,r}(s))^2\ 1_{(a_r^y|\Delta M^y_{\alpha,r}(s)|>\varepsilon)}$$

$$= P_{\alpha,r}\ \sum_{s\le t} a_r^{y\,2}(\Delta\tilde{N}_r^y(s))^2\ \dfrac{1}{(\tilde{Z}_r^y(s-))^2}\ 1\left(\tilde{Z}_r^y(s-)>0,\ \dfrac{\Delta\tilde{N}_r^y(s)}{\tilde{Z}_r^y(s-)} > \dfrac{\varepsilon}{a_r^y}\right).$$

But $\Delta\tilde{N}_r^y(s)$ is 0 or 1, so this equals

$$P_{\alpha,r}\ \int_{(0,t]} \tilde{N}_r^y(ds)\ \dfrac{a_r^{y\,2}}{(\tilde{Z}_r^y(s-))^2}\ 1\left(0<\tilde{Z}_r^y(s-)<\dfrac{a_r^y}{\varepsilon}\right)$$

$$= P_{\alpha,r}\ \int_0^t ds\ \alpha^y(s)\ \dfrac{a_r^{y\,2}}{\tilde{Z}_r^y(s)}\ 1\left(0<\tilde{Z}_r^y(s)<\dfrac{a_r^y}{\varepsilon}\right)$$

using Theorem 3.2.8(i) for the last equality. Thus condition (a) in the statement of Theorem 2.2 matches (a) of Theorem 1.3 and the proof is complete. $\quad\blacksquare$

In applications, the verification of conditions (a) and (b) in Theorem 2.2, can be quite difficult. When checking (a), it is useful to know that (a) is satisfied provided

(2.3)
$$\lim_{r\to\infty} \sup_{s\le t} P_{\alpha,r} \, X_r^y(s) 1_{(X_r^y(s) > a_r^y \epsilon)} = 0$$

where

$$X_r^y(s) = \frac{a_r^{y^2}}{\tilde{Z}_r^y(s)} 1_{(\tilde{Z}_r^y(s) > 0)} \ ,$$

as is seen rewriting (a) by changing the order of integration, as

(2.4)
$$\lim_{r\to\infty} \int_0^t ds \; \alpha^y(s) P_{\alpha,r} \, X_r^y(s) 1_{(X_r^y(s) > a_r^y \epsilon)} = 0 \ ,$$

and then using (2.3) and dominated convergence to establish (2.4).

Since $a_r^y \to \infty$, (2.3) states that asymptotically X_r^y can only be large with small probability, which by the definition of X_r^y amounts to saying that $\tilde{Z}_r^y$ must be large with high probability.

We shall now see how Theorem 2.2 can be applied to some of the particular models we have discussed previously.

For these applications the following result will prove useful.

2.5. <u>Lemma</u>. Suppose the random variable X defined on the probability space $(\Omega, A, \mathbb{P})$ is binomial (n, p), i.e.

$$\mathbb{P}(X = x) = \binom{n}{x} p^x (1-p)^{n-x} \qquad (x = 0, 1, \cdots, n) \ ,$$

where $0 \le p \le 1$. Then for any a, $0 \le a \le n$,

$$\mathbb{P}(X > a) \le e^{-a} (pe + 1-p)^n \ ,$$

$$\mathbb{P}(X < a) \le e^a (pe^{-1} + 1-p)^n \ .$$

<u>Proof</u>. For the first inequality, write

$$\mathbb{P}(X > a) = \mathbb{P}(e^X > e^a)$$

$$\le e^{-a} \, \mathbb{P} \, e^X = e^{-a} (pe + 1-p)^n \ ,$$

and for the second

$$\mathbb{P}\,(X < a) \;=\; \mathbb{P}\,(e^{-X} > e^{-a})$$

$$\leq\; e^a\,\mathbb{P}\,e^{-X} \;=\; e^a(pe^{-1} + 1-p)^n\,. \qquad \blacksquare$$

2.6. **Example**. The simplest product model comes by considering i.i.d. lifetimes, cf. Example 4.3.7. With μ the intensity for the survival distribution we have that

$$\beta^*_{\mu,r}(t) \;=\; \int_0^t ds\ \mu(s)\ 1_{(\tilde{N}_r(s) < r)}$$

is estimated by

$$\hat{\beta}_r(t) \;=\; \int_{(0,t]} \tilde{N}_r(ds)\ \frac{1}{r-\tilde{N}_r(s-)}\ 1_{(\tilde{N}_r(s-) < r)}\,.$$

Now take $a_r = \sqrt{r}$ and look at

$$\sqrt{r}\ M_{\mu,r}(t) \;=\; \sqrt{r}\ (\hat{\beta}_r(t) - \beta^*_{\mu,r}(t))$$

which is a $P_{\mu,r}$-martingale with

$$(2.7) \qquad < \sqrt{r}\ M_{\mu,r} >_t \;=\; \int_0^t ds\ \mu(s)\ \frac{r}{r-\tilde{N}_r(s)}\ 1_{(\tilde{N}_r(s) < r)}\,.$$

We shall show that $\sqrt{r}\ M_{\mu,r}$ <u>converges in distribution to the Gauss-Φ process where</u> Φ <u>is given by (2.9) below.</u>

If $G_\mu(t) = \exp(-\int_0^t \mu)$ is the survivor function with intensity μ and $F_\mu = 1-G_\mu$, then with respect to $P_{\mu,r}$, $\tilde{N}_r(s)$ is binomial $(r, F_\mu(s))$, so by the strong law of large nembers

$$(2.8) \qquad \frac{1}{r}\ \tilde{N}_r(s) \;\rightarrow\; F_\mu(s)$$

almost surely. (Formally, to make this statement, all processes $\tilde{N}_r$ must be defined on the same probability space, which they are not. Things can be made rigorous by embedding everything into an infinite product space, but we shall not worry about this).

With $\mu \in A$, $G_\mu(t) > 0$ for all $t \geq 0$, so by (2.8), for any fixed

$t \geq 0$, almost surely for $r \geq r_0$ sufficiently large (with r_0 random), the sequence $(\frac{1}{r} \tilde{N}_r(t))$ is bounded above by a number < 1, and since $\tilde{N}_r(s) \leq \tilde{N}_r(t)$ for $s \leq t$ also

$$\sup_{r \geq r_0} \sup_{s \leq t} \frac{1}{r} \tilde{N}_r(s) < 1$$

almost surely. But then

$$\frac{r}{r - \tilde{N}_r(s)} \, 1 \, (\tilde{N}_r(s) < r) \quad \rightarrow \quad \frac{1}{G_\mu(s)}$$

almost surely and

$$\sup_{r \geq r_0} \sup_{s \leq t} \frac{r}{r - \tilde{N}_r(s)} \, 1 \, (\tilde{N}_r(s) < r) \quad < \infty$$

almost surely, so using dominated convergence in (2.7) for almost every realisation of the sequence $(\tilde{N}_r)_{r \geq 1}$, it emerges that

$$<\sqrt{r} \, M_{\mu,r}>_t \quad \rightarrow \quad \int_0^t ds \, \mu(s) \, \frac{1}{G_\mu(s)}$$

$$= \int_0^t ds \, \mu(s) e^{\int_0^s \mu}$$

$$= e^{\int_0^t \mu} - 1 = \frac{F_\mu(t)}{G_\mu(t)}$$

almost surely. Since we then also have convergence in probability, it follows that (b) of Theorem 2.2 is satisfied with

(2.9) $$\omega(t) = \frac{F_\mu(t)}{G_\mu(t)} \, .$$

To check (a), it is by (2.3) enough to show that

(2.10) $$\lim_{r \to \infty} \sup_{s \leq t} P_{\mu,r} \frac{r}{r - \tilde{N}_r(s)} \, 1 \, (\tilde{N}_r(s) < r, \, \frac{r}{r - \tilde{N}_r(s)} > \sqrt{r} \, \varepsilon) = 0 \, .$$

Here the indicator function equals

$$\left(r - \frac{\sqrt{r}}{\epsilon} < \tilde{N}_r(s) < r \right)^{1} ,$$

so by the crude estimate $\dfrac{r}{r - \tilde{N}_r(s)} \le r$ on $(\tilde{N}_r(s) < r)$, we see that (2.10) follows from

$$(2.11) \qquad \limsup_{\substack{r \to \infty \\ s \le t}} r \, P_{\mu,r}(\tilde{N}_r(s) > r - \frac{\sqrt{r}}{\epsilon}) = 0 .$$

But by Lemma 2.5

$$(2.12) \qquad r \, P_{\mu,r}(\tilde{N}_r(s) > r - \frac{\sqrt{r}}{\epsilon}) \le r \, \exp(-r + \frac{\sqrt{r}}{\epsilon}) \, (F_\mu(s)e + G_\mu(s))^r ,$$

and since for $s \le t$

$$F_\mu(s)e + G_\mu(s) \le F_\mu(t)e + G_\mu(t) < e ,$$

the right-hand side of (2.12) tends to 0 as $r \to \infty$ uniformly in $s \le t$, and (2.11) follows.

As usual we have assumed above that $\mu \in A$, in particular $\int_0^t \mu < \infty$ for all t. If one considers a μ such that G_μ has finite termination point $t^\dagger$, the arguments above are still valid as long as $t < t^\dagger$.

So we know now that $\sqrt{r} \, (\hat{\beta}_r - \beta^*_{\mu,r})$ converges in distribution to the Gauss-Φ process. As we shall now see, $\sqrt{r} \, (\hat{\beta}_r - \beta^{**}_\mu)$ has the same limit distribution, where β^{**}_μ is the true integrated intensity

$$\beta^{**}_\mu(t) = \int_0^t ds \, \mu(s) .$$

For this it is enough to show that

$$(2.13) \qquad \sqrt{r} \, \sup_{s \le t} \left| \beta^*_{\mu,r}(s) - \beta^{**}_\mu(s) \right|$$

converges in probability to 0 for every t, and this is clear since the quantity (2.13) is $\le \sqrt{r} \int_0^t \mu \, 1_{(\tilde{N}_r(t) = r)}$ and $\sqrt{r} \, P_{\mu,r}(\tilde{N}_r(t) = r) \to 0$.

We saw in Example 4.3.7, that $\hat{\beta}_r$ is the (discrete) intensity for the empirical survivor function $\hat{G}_r$, which is then the product-limit estimator of G_μ. We shall in the next section derive the asymptotic distribution of $\sqrt{r}\,(\hat{G}_r - G_\mu)$ from Theorem 1.3, and only here point out another asymptotic result concerning $\hat{F}_r = 1-\hat{G}_r$, which is not very useful, but immediate to get. Since $\hat{F}_r(t) = \frac{1}{r}\,\tilde{N}_r(t)$,

$$\check{M}_{\mu,r}(t) = \hat{F}_r(t) - \frac{1}{r}\int_0^t ds\,\mu(s)\,(r-\tilde{N}_r(s))^+$$

is a $P_{\mu,r}$-martingale with

$$<\check{M}_{\mu,r}>_t = \frac{1}{r^2}\int_0^t ds\,\mu(s)\,(r-\tilde{N}_r(s))^+ \ .$$

With arguments like the ones used above, it follows from Theorem 1.3 that $\sqrt{r}\,\check{M}_{\mu,r}$ converges in distribution to the Gauss-$\check{\Phi}$ process, where $\check{\Phi}(t) = F_\mu(t)$. ❙

2.14.Example. Consider the product model from Example 4.3.8 with i.i.d. lifetimes and fixed censoring times. With $\mu \in A$ the intensity for the survival distribution, and $u_1, u_2, \cdots$ the sequence of censoring times, the intensity for $P_{\mu,r}$, the process for the first r individuals, is $\lambda_- = (\lambda_-^i)_{1\le i\le r}$ where

$$\lambda_{t-}^i = \mu(t-)1_{(i\in R_r(t-))}$$

with $R_r(t-) = \{i: 1 \le i \le r, \tau_1^i \ge t, u_i \ge t\}$ the set of individuals among $1,\cdots,r$ at risk immediately before t.

For this model we have that

$$\beta_{\mu,r}^*(t) = \int_0^t ds\,\mu(s)1_{(|R_r(s)|>0)}$$

is estimated by

$$\hat{\beta}_r(t) = \int_{(0,t]} \tilde{N}_r(ds)\,\frac{1}{|R_r(s-)|}\,1_{(|R_r(s-)|>0)}$$

with $\tilde{N}_r(s)$ the number of individuals among $1,\cdots,r$ observed to have died during $[0,s]$. Furthermore, $M_{\mu,r} = \hat{\beta}_r - \beta_{\mu,r}^*$ is a $P_{\mu,r}$-martingale with

$$< M_{\mu,r} >_t = \int_0^t ds \, \mu(s) \, \frac{1}{|R_r(s)|} \, 1 \, (|R_r(s)| > 0) \ .$$

To get asymptotic results we shall need a condition on the sequence (u_i). We shall assume that for all $s \geq 0$

(2.15)
$$\frac{1}{r} n_r(s) \underset{r \to \infty}{\to} \varphi(s) \ ,$$

where $n_r(s) = \#\{i: 1 \leq i \leq r, u_i > s\}$ and $\varphi: [0,\infty) \to [0,1]$ is non-increasing with $\varphi(0) = 1$.

To get results valid for the entire time axis, we shall also assume that $\varphi(s) > 0$ for all $s \geq 0$. This condition may be dropped, but
then the results below hold only strictly to the left of $s^\dagger = \inf\{s: \varphi(s) = 0\}$.

If one considers a model with random censoring times $U_1, U_2, \cdots$
such that the U_i are i.i.d. and independent of the lifetimes, then
(2.15) holds almost surely by the strong law of large numbers, and the
results below apply to this model when conditioning on the U_i.

We shall now show that as $r \to \infty$, $\sqrt{r} \, (\hat{\beta}_r - \beta^*_{\mu,r})$ <u>converges in dis-</u>
<u>tribution to the Gauss-Φ process with</u>

(2.16)
$$\Phi(t) = \int_0^t ds \, \mu(s) \, \frac{1}{\varphi(s) G_\mu(s)} \ ,$$

writing $G_\mu(s) = 1 - F_\mu(s) = e^{-\int_0^s \mu}$.

The important fact needed for the verification of the conditions
from Theorem 2.2 is that

$$|R_r(s)| = \sum_{\substack{i=1 \\ u_i > s}}^{r} 1 \, (\tau_1^i > s)$$

is binomial $(n_r(s), G_\mu(s))$.

Now fix $t > 0$. For $s \leq t$, by the strong law of large numbers
and (2.15)

$$\frac{1}{r} |R_r(s)| = \frac{n_r(s)}{r} \, \frac{1}{n_r(s)} \, |R_r(s)| \to \varphi(s) G_\mu(s)$$

almost surely, the limit exceeding $\varphi(t) G_\mu(t) > 0$, and since

$$< \sqrt{r}\, M_{\mu,r} >_t \; = \; \int_0^t ds\; \mu(s)\; \frac{r}{|R_r(s)|}\; 1_{(|R_r(s)| > 0)}\; ,$$

we can use dominated convergence to obtain

$$< \sqrt{r}\, M_{\mu,r} >_t \; \rightarrow \; \int_0^t ds\; \mu(s)\; \frac{1}{\varphi(s) G_\mu(s)}\; ,$$

almost surely, so that (b) of Theorem 2.2 is satisfied with Φ given by (2.16).

To verify (a) we use (2.3) which becomes

$$\sup_{s \leq t} P_{\mu,r}\; \frac{r}{|R_r(s)|}\; 1_{(0 < |R_r(s)| < \frac{\sqrt{r}}{\epsilon})} \; \rightarrow \; 0 \; ,$$

and therefore will follow if

$$\lim_{r \rightarrow \infty} \sup_{s \leq t} r\, P_{\mu,r}(|R_r(s)| < \frac{\sqrt{r}}{\epsilon}) \; = \; 0 \; .$$

But by Lemma 2.5,

$$r\, P_{\mu,r}(|R_r(s)| < \frac{\sqrt{r}}{\epsilon}) \; \leq \; r e^{\frac{\sqrt{r}}{\epsilon}} \; (G_\mu(s) e^{-1} + F_\mu(s))^{n_r(s)} \; ,$$

and here for $s \leq t$

$$G_\mu(s) e^{-1} + F_\mu(s) \; \leq \; G_\mu(t) e^{-1} + F_\mu(t) < 1$$

so that because $n_r(s) \geq n_r(t)$

$$\sup_{s \leq t} r\, P_{\mu,r}(|R_r(s)| < \frac{\sqrt{r}}{\epsilon}) \; \leq \; r e^{\frac{\sqrt{r}}{\epsilon}} \; (G_\mu(t) e^{-1} + F_\mu(t))^{n_r(t)} \; ,$$

which by (2.15) is seen to converge to 0, completing the proof.

In analogy with Example 2.6 one can also show that $\sqrt{r}\,(\hat{\beta}_r - \beta_\mu^{**})$, where $\beta_\mu^{**}(t) = \int_0^t \mu$, converges in distribution to the same Gauss-Φ process. ∎

2.17. Example. We shall discuss the asymptotic behavior of the estimators of the integrated transition intensities in the product Markov chain model treated in Section 4.4.

With the notation from there we have that for $(i,j) \in E$ $(i \neq j \in S)$,

$$\beta_{\mu,r}^{*ij}(t) = \int_0^t ds\ \alpha_{ij}(s) 1_{(S_r^i(s) > 0)}$$

is estimated by

$$\hat{\beta}_r^{ij}(t) = \int_{(0,t]} \hat{N}_r^{ij}(ds)\ \frac{1}{S_r^i(s-)}\ 1_{(S_r^i(s-) > 0)} \ .$$

Furthermore, the processes $(M_{\alpha,r}^{ij})_{(i,j)\in E} = (\hat{\beta}_r^{ij} - \beta_{\mu,r}^{*ij})_{(i,j)\in E}$ are orthogonal $P_{\alpha,r}$-martingales with

(2.18) $$< M_{\alpha,r}^{ij} >_t = \int_0^t ds\ \alpha_{ij}(s)\ \frac{1}{S_r^i(s)}\ 1_{(S_r^i(s) > 0)} \ .$$

Writing i_ℓ for the initial state of the ℓ'th chain, we shall assume that for every $i \in S$

(2.19) $$\frac{1}{r}\ n_r(i) \underset{r\to\infty}{\to} p_i$$

where $n_r(i) = \# \{\ell: 1 \le \ell \le r\ ;\ i_\ell = i\}$. Of course $p_i \ge 0$ with $\Sigma\ p_i = 1$. This condition is satisfied if one takes the model of independent Markov chains with common transition probabilities and common initial distribution given by the point probabilities p_i , and then condition on the initial states.

We shall show, that if all $p_i > 0$, the sequence $(\sqrt{r}(\hat{\beta}_r^{ij} - \beta_{\alpha,r}^{*ij}))_{(i,j)\in E}$ of multidimensional martingales <u>converges in distribution to the Gauss-Φ process with</u> $\Phi = (\Phi^{ij})_{(i,j)\in E}$ <u>given by</u>

(2.20) $$\Phi^{ij}(t) = \int_0^t ds\ \frac{\alpha_{ij}(s)}{p_i(s)} \ ,$$

writing $p_i(s) = \Sigma_j\ p_j p_{ji}(0,s)$ for the probability that a chain with initial distribution (p_j) and transitions $(p_{ij}(s,t))$ is in state i at time s . Here $p_{ij}(s,t)$ is determined from the α_{ij} by (4.4.5),

(4.4.6), (4.4.7). The assumptions guarentee that the integral in (2.20) converges:

$$(2.21) \qquad P_i(s) \geq P_i \, P_{ii}^{(0)}(0,s) = P_i \, \exp\left(-\int_0^s \mu_i\right)$$

$$\geq P_i \, \exp\left(-\int_0^t \mu_i\right) > 0$$

for $s \leq t$.

We first verify condition (b) of Theorem 2.2. Because of (2.19) and the strong law of large numbers, for any $i \in S$

$$\frac{1}{r} S_r^i(s) = \sum_{j \in S} \frac{n_r(j)}{r} \, \frac{1}{n_r(j)} \sum_{\substack{\ell=1 \\ i_\ell=j}}^{r} 1 \, (J_s^\ell = i)$$

$$\to \sum_{j \in S} P_j \, P_{ji}(0,s) = P_i(s)$$

almost surely. To justify by dominated convergence, that this limit may be performed under the integration sign in (2.18), use the bound

$$\inf_{s \leq t} \frac{1}{r} S_r^i(s) \geq \frac{1}{r} \sum_{\ell=1}^{r} 1 \, (J_s^\ell = i \text{ for all } s \leq t),$$

together with the fact that by the strong law of large numbers, the average on the right converges almost surely to

$$P_i \, P_{ii}^{(0)}(0,t) = P_i \, \exp\left(-\int_0^t \mu_i\right) > 0 \ .$$

It follows thus that

$$< \sqrt{r} \, M_{\alpha,r}^{ij} >_t \ \to \ \Phi^{ij}(t)$$

almost surely, and (b) is verified.

To establish (a) of Theorem 2.2, using (2.3) we must show that for all $(i,j) \in E$, $t \geq 0$

$$\sup_{s \leq t} P_{\mu,r} \frac{r}{S_r^i(s)} \, 1 \, (0 < S_r^i(s) < \frac{\sqrt{r}}{\varepsilon} \,) \ \to 0 \ ,$$

and for this it is enough that

(2.22)
$$\sup_{s \le t} r\, P_{\mu,r}(S^i_r(s) < \frac{\sqrt{r}}{\varepsilon}) \to 0 .$$

But (and this is a variant of Lemma 2.5)

(2.23)
$$r\, P_{\mu,r}(S^i_r(s) < \frac{\sqrt{r}}{\varepsilon}) = r\, P_{\mu,r}\left(\exp\left(-S^i_r(s)\right) > e^{-\frac{\sqrt{r}}{\varepsilon}}\right)$$

$$\le r\, e^{\frac{\sqrt{r}}{\varepsilon}}\, P_{\mu,r}\left(\exp\left(-S^i_r(s)\right)\right) ,$$

where for $s \le t$

$$P_{\mu,r}\,\exp\left(-S^i_r(s)\right) = \prod_{\ell=1}^{r} P_{\mu,r}\,\exp\left(-1_{(J^\ell_s = i)}\right)$$

$$\le \prod_{\substack{\ell=1 \\ i_\ell = i}}^{r} P_{\mu,r}\,\exp\left(-1_{(J^\ell_s = i)}\right)$$

$$= (p_{ii}(0,s)e^{-1} + 1 - p_{ii}(0,s))^{n_r(i)}$$

$$\le (e^{-1} + (1-e^{-1})(1 - e^{-\int_0^t \mu_i}))^{n_r(i)}$$

using $p_{ii}(0,s) \ge p^{(0)}_{ii}(0,s)$ and (2.21) for the last step. By (2.19) and the assumption $p_i > 0$, this is more than good enough to imply (2.22) via (2.23).

The parameter $\mu_i = \sum_{j \ne i} \alpha_{ij}$ is the intensity for the waiting time distribution in state i. From Section 4.4 we have that

$$\beta^{*i\cdot}_{\mu,r}(t) = \int_0^t ds\, \mu_i(s) 1_{(S^i_s > 0)}$$

is estimated by

$$\hat{\beta}^{i\cdot}_r(t) = \int_{(0,t]} \tilde{N}^{i\cdot}(ds)\, \frac{1}{S^i_{s-}}\, 1_{(S^i_{s-} > 0)} .$$

From the limit result we have just shown, it follows immediately that $(\sqrt{r}(\hat{\beta}^{i\cdot}_r - \beta^{*i\cdot}_{\mu,r}))_{i \in S}$ converges in distribution to the Gauss-$\Phi^{\cdot}$ process with $\Phi^{\cdot} = (\Phi^{i\cdot})_{i \in S}$ given by

$$\Phi^{i\cdot}(t) = \sum_{j \ne i} \Phi^{ij}(t) = \int_0^t ds\, \frac{\mu_i(s)}{p_i(s)} . \qquad\blacksquare$$

Under the conditions of Theorem 2.2, we have that $(a_r^y(\hat{\beta}_r^y - \beta_{\alpha,r}^{*y}))_{y\in E}$ is approximately Gauss-Φ with independent increments. Since however typically, the Φ^y will depend on the unknown α^y, to use the theorem in practice we must estimate something like the asymptotic variance of $\hat{\beta}_r^y$. But $\Phi^y(t)$ is the limit in probability of

$$\omega_{\alpha,r}^{y2}(t) = \int_0^t ds \ \alpha^y(s) \ \frac{a_r^{y2}}{\tilde{Z}_r^y(s)} 1 \ (\tilde{Z}_r^y(s) > 0) \quad,$$

so we shall find an estimator for this and then use that as our guess of $\Phi^y(t)$. We define (cf. Proposition 4.3.6)

$$\hat{\omega}_r^{y2}(t) = \int_{(0,t]} \tilde{N}_r^y(ds) \ \frac{a_r^{y2}}{(\tilde{Z}_r^y(s-))^2} 1 \ (\tilde{Z}_r^y(s-) > 0) \quad.$$

2.24. Proposition. (a) The processes $\hat{\omega}_r^{y2} - \omega_{\alpha,r}^{y2}, (y\in E)$ are orthogonal $P_{\alpha,r}$-martingales with

$$<\hat{\omega}_r^{y2} - \omega_{\alpha,r}^{y2}>_t = \int_0^t ds \ \alpha^y(s) \frac{a_r^{y4}}{(\tilde{Z}_r^y(s))^3} 1 \ (\tilde{Z}_r^y(s) > 0) \quad.$$

(b) If the conditions of Theorem 2.2 are satisfied, and if for $t \geq 0$

$$\lim_{r\to\infty} a_r^{y4} \int_0^t ds \ \alpha^y(s) P_{\alpha,r} \frac{1}{(\tilde{Z}_r^y(s))^3} 1 \ (\tilde{Z}_r^y(s) > 0) = 0 \ ,$$

then $\hat{\omega}_r^{y2}$ is a consistent estimator of Φ^y: for every $y \in E$, $t \geq 0$, $\varepsilon > 0$

$$\lim_{r\to\infty} P_{\alpha,r}(|\hat{\omega}_r^{y2}(t) - \Phi^y(t)| > \varepsilon) = 0 \ .$$

Proof. (a) follows from Theorem 3.2.8. As for (b) just observe that

$$P_{\alpha,r}(|\omega_r^{y2}(t) - \Phi^y(t)| > \varepsilon)$$

$$\leq P_{\alpha,r}(|\hat{\omega}_r^{y2}(t) - \omega_{\alpha,r}^{y2}(t)| > \frac{\varepsilon}{2}) + P_{\alpha,r}(|\omega_{\alpha,r}^{y2}(t) - \Phi^y(t)| > \frac{\varepsilon}{2}) \ ,$$

where the second term tends to 0 by condition (b), Theorem 2.2, while the first is dominated by

$$\frac{4}{\varepsilon^2} P_{\alpha,r}(\hat{\omega}_r^{y^2}(t) - \omega_{\alpha,r}^{y^2}(t))^2 = \frac{4}{\varepsilon^2} P_{\alpha,r} < \hat{\omega}_r^{y^2} - \omega_{\alpha,r}^{y^2} >_t \ ,$$

and the assumption in part (b) of the proposition is precisely that

this $\to 0$　　　　　　　　　　　　　　　　　　　　　　　　　　　　|

Condition (b) is satisfied in particular if

$$\lim_{r\to\infty} \sup_{s\le t} P_{\alpha,r} \frac{a_r^{y^4}}{\left(\hat{z}_r^y(s)\right)^3} 1_{(\hat{z}_r^y(s) > 0)} = 0 \ .$$

By arguments similar to the one to be presented in Example 2.28 below, this condition can be verified for the cases discussed in Examples 2.6, 2.14, 2.17. The details are left to the reader.

Under the conditions of Theorem 2.2, it follows in particular, because $a_r^y \to \infty$, that $\hat{\beta}_r^y(t)$ is a consistent estimator for $\beta_{\alpha,r}^{*y}(t)$, in the sense that the difference $\hat{\beta}_r^y(t) - \beta_{\alpha,r}^{*y}(t)$ tends to 0 in probability. Under a different condition (not assuming (a) or (b) of Theorem 2.2), this can be improved to a form of uniform consistency as we shall now see.

2.25. **Proposition.** Suppose that

(2.26)　　　$\lim_{r\to\infty} \int_0^t ds \ a^y(s) P_{\alpha,r} \frac{1}{\hat{z}_r^y(s)} 1_{(\hat{z}_r^y(s) > 0)} = 0$　$(y \in E, t \ge 0)$.

Then for every $y \in E, t \ge 0$

$$\lim_{r\to\infty} P_{\alpha,r} \sup_{s\le t} \left(\hat{\beta}_r^y(s) - \beta_{\alpha,r}^{*y}(s)\right)^2 = 0 \ .$$

Proof. By Doob's martingale inequality

$$P_{\alpha,r} \sup_{s\le t} M_{\alpha,r}^{y^2}(s) \le 4 P_{\alpha,r} M_{\alpha,r}^{y^2}(t) \ ,$$

where as before $M_{\alpha,r}^y = \hat{\beta}_r^y - \beta_{\alpha,r}^{*y}$.　But by (2.1)

$$P_{\alpha,r} \, M^{y}{}^{2}_{\alpha,r}(t) = P_{\alpha,r} < M^{y}_{\alpha,r} > t$$

$$= P_{\alpha,r} \int_0^t ds \, \alpha^y(s) \, \frac{1}{\tilde{Z}^y_r(s)} \, 1 \, (\tilde{Z}^y_r(s) > 0) \quad .$$

Now interchange the order of integration. |

Using that $\alpha^y \in A$ and dominated convergence, it is seen that (2.26) is satisfied if

(2.27) $\lim\limits_{\substack{r \to \infty \\ s \leq t}} \sup P_{\alpha,r} \, \frac{1}{\tilde{Z}^y_r(s)} \, 1 \, (\tilde{Z}^y_r(s) > 0) = 0$

for all $y \in E$, $t \geq 0$.

2.28. <u>Example</u>. To check that Proposition 2.25 is valid for the case discussed in Example 2.6, we verify (2.27). The expectation is

$$P_{\mu,r} \, \frac{1}{r - \bar{N}_r(s)} \, 1 \, (\check{N}_r(s) < r)$$

$$= P_{\mu,r} \, \frac{1}{r - \bar{N}_r(s)} \, 1 \, (\check{N}_r(s) \leq ra) + P_{\mu,r} \, \frac{1}{r - \bar{N}_r(s)} \, 1 \, (ra < \check{N}_r(s) < r)$$

for any a, $0 < a < 1$. The first term is $\leq r^{-1}(1-a)^{-1}$, and the second is dominated by (for $s \leq t$)

$$P_{\mu,r}(\check{N}_r(s) > ra) \leq e^{-ra}(F_{\mu}(s) e + G_{\mu}(s))^r$$

$$\leq e^{-ra}(F_{\mu}(t) e + G_{\mu}(t))^r$$

by Lemma 2.5. But $F_{\mu}(t) e + G_{\mu}(t) = c_t < e$, and it is seen that (2.27) holds if we choose a such that $\log c_t < a < 1$ □

By similar methods it can be shown that Proposition 2.25 is valid for the cases in Examples 2.14 and 2.17.

5.3. <u>Asymptotic distributions of product-limit estimators.</u>

We shall still consider the Aalen product model as described in
the previous section.

In many applications, the parameters α^y are intensities for unknown survivor functions

$$G_{\alpha^y}(t) = \exp\left(-\int_0^t \alpha^y\right)$$

(cf. Examples 2.6 and 2.14), and we have previously in Examples 4.3.7
and 4.3.8 argued that a natural estimator for G_{α^y} is the product-limit
estimator

(3.1)
$$\hat{G}_r^y(t) = \prod_{s \leq t} (1 - \Delta\hat{\beta}_r^y(s))$$

$$= \prod_{s \leq t} \left(1 - \frac{\Delta\widetilde{N}_r^y(s)}{\widetilde{Z}_r^y(s-)} \, 1_{(\widetilde{Z}_r^y(s-) > 0)}\right).$$

Even in the general case, where the G_{α^y} may not have interpretations
as survivor functions one may of course still consider $\hat{G}_r^y$ as estimator of G_{α^y}, and we shall now discuss the asymptotic behavior of this
estimator.

For two reasons, we cannot expect $\hat{G}_r^y(t)$ to be a reasonable estimator of $G_{\alpha^y}(t)$ for all values of t: firstly, because $0 \leq G_{\alpha^y} \leq 1$
it is desirable that also $0 \leq \hat{G}_r^y \leq 1$, wherefore negative factors must
not be allowed in the product in (3.1); secondly, we saw in Section 4.4
a situation where, because $\hat{\beta}_r^y(t)$ estimates $\alpha^y(t)$ only at t with
$\widetilde{Z}_r^y(t) > 0$, the natural estimator for G_{α^y} was a family of survivor
functions, each supporting a probability on one of a number of disjoint
intervals.

Therefore, defining

$$\tau_r^{*y} = \inf\{t \geq 0 : \widetilde{Z}_r^y(t) < 1\},$$

we shall only estimate G_{α^y} on $[0, \tau_r^{*y}]$.

In all the examples of the previous section, $\tilde{Z}_r^y \geq 0$ was integer-valued so that the problem of $\hat{G}_r^y$ becoming negative did not arise. If $\tilde{Z}_r^y \geq 0$ is always an integer

$$\tau_r^{*y} = \inf\{t \geq 0 : \tilde{Z}_r^y(t) = 0\}.$$

It is not too difficult to show, that τ_r^{*y} is measurable, and because the process $\tilde{Z}_r^y$ is adapted, it is then clear that τ_r^{*y} is a stopping time.

The definition of τ_r^{*y} implies that

(3.2) $$\tilde{Z}_r^y(\tau_r^{*y} -) \geq 1 \qquad \text{on } (\tau_r^{*y} < \infty)$$

for all $y \in E, r \geq 1$.

For $y \in E, r \geq 1$, introduce the process $Q_r^y = (Q_r^y(t))_{t \geq 0}$ by

$$Q_r^y(t) = 1 - \frac{\hat{G}_r^y(t \wedge \tau_r^{*y})}{G_{\alpha y}(t \wedge \tau_r^{*y})}.$$

Of course Q_r^y has right-continuous, left-limit paths, and $Q_r^y(0) = 0$.

3.3. Lemma. For all $y \in E, t \geq 0$

(3.4) $$Q_r^y(t) = \int_{(0,t]} \tilde{M}_r^y(ds) \frac{\hat{G}_r^y(s-)}{G_{\alpha y}(s)} \frac{1}{\tilde{Z}_r^y(s-)} 1_{(\tau_r^{*y} \geq s)},$$

where $\tilde{M}_r^y$ is the $P_{\alpha,r}$-martingale

$$\tilde{M}_r^y(t) = \tilde{N}_r^y(t) - \int_0^t ds \, \alpha^y(s) \tilde{Z}_r^y(s).$$

In particular, for $y \in E$ the processes Q_r^y, $(y \in E)$, are orthogonal $P_{\alpha,r}$-martingales with

$$< Q_r^y >_t = \int_0^t ds \, \alpha^y(s) \left(\frac{\hat{G}_r^y(s)}{G_{\alpha y}(s)}\right)^2 \frac{1}{\tilde{Z}_r^y(s)} 1_{(\tau_r^{*y} > s)}.$$

<u>Remark</u>. The stochastic integral on the right of (3.4) extends from 0 to $t \wedge \tau_r^{*y}$, the latter point included. The integral is well defined because of (3.2).

<u>Proof</u>. The random part of the integrand on the right of (3.4) is left-continuous, so the integrand is predictable. Therefore, once (3.4) is proved, everything else follows from Theorem 3.2.8.

For the proof of (3.4) notice that both sides are right-continuous in t, taking the values 0 for $t = 0$. So it is enough to show that the increments of both sides from $\tilde{\tau}_k^y$ to $t \leq \tilde{\tau}_{k+1}^y \wedge \tau_r^{*y}$ are the same, where $\tilde{\tau}_k^y$ is the time of the k'th jump of $\tilde{N}_r^y$. Thus we shall show that on $(\tilde{\tau}_k^y < \tau_r^{*y})$

$$(3.5) \qquad \frac{\hat{G}(\tau_k)}{G(\tau_k)} - \frac{\hat{G}(t)}{G(t)} = \int_{(\tau_k, t]} M(ds) \; \frac{\hat{G}(s-)}{G(s)} \; \frac{1}{Z(s-)}$$

for t such that $\tau_k < t < \tau_{k+1}$, $t \leq \tau^*$, and that

$$(3.6) \qquad \frac{\hat{G}(\tau_k)}{G(\tau_k)} - \frac{\hat{G}(\tau_{k+1})}{G(\tau_{k+1})} = \int_{(\tau_k, \tau_{k+1}]} M(ds) \; \frac{\hat{G}(s-)}{G(s)} \; \frac{1}{Z(s-)}$$

on $(\tau_{k+1} < \infty, \tau_{k+1} \leq \tau^*)$. (To lighten the notation, we have omitted various subscripts and other adornments).

Using the definition of M, the integral in (3.5) becomes

$$- \int_{\tau_k}^{t} ds \; \alpha(s) \; \frac{\hat{G}(s)}{G(s)}$$

because $\tilde{N}_r^y$ has no jumps on $(\tau_k, t]$. But $\hat{G}$ is constant on $(\tau_k, t]$, so this reduces to

$$- \hat{G}(\tau_k) \int_{\tau_k}^{t} ds \; \alpha(s) \; \frac{1}{G(s)} = - \hat{G}(\tau_k) \left(\frac{1}{G(t)} - \frac{1}{G(\tau_k)} \right)$$

and (3.5) follows since $\hat{G}(\tau_k) = \hat{G}(t)$.

As for (3.6), since $\tilde{N}_r^y$ jumps at τ_{k+1}, the integral there becomes

$$\frac{\hat{G}(\tau_{k+1}-)}{G(\tau_{k+1})} \frac{1}{Z(\tau_{k+1}-)} - \int_{\tau_k}^{\tau_{k+1}} ds \; \alpha(s) \frac{\hat{G}(s)}{G(s)} \; .$$

But $\hat{G}$ is constant on (τ_k, τ_{k+1}), and by (3.1)

$$\hat{G}(\tau_{k+1}) = \hat{G}(\tau_k) \left(1 - \frac{1}{Z(\tau_{k+1}-)}\right) ,$$

so this reduces to

$$\frac{\hat{G}(\tau_k)}{G(\tau_{k+1})} \frac{1}{Z(\tau_{k+1}-)} - \hat{G}(\tau_k)\left(\frac{1}{G(\tau_{k+1})} - \frac{1}{G(\tau_k)}\right)$$

$$= \frac{\hat{G}(\tau_k)}{G(\tau_k)} - \frac{\hat{G}(\tau_k)}{G(\tau_{k+1})}\left(1 - \frac{1}{Z(\tau_{k+1}-)}\right)$$

$$= \frac{\hat{G}(\tau_k)}{G(\tau_k)} - \frac{\hat{G}(\tau_{k+1})}{G(\tau_{k+1})}$$

as desired.   ▌

The representation (3.4) permits the use of Theorem 1.3 in evalua-
ting the asymptotic distribution of Q_r^y, and from this it is easy to
arrive at the following result, in which the limit process is Gaussian,
but does not have independent increments. As usual we write $F_\alpha^y = 1 - G_\alpha^y$,
$\hat{F}_r^y = 1 - \hat{G}_r^y$.

3.7. _Theorem_. Suppose there exists a family $\Phi_\alpha = (\Phi_\alpha^y)_{y \in E}$ of non-
decreasing, continuous functions $\Phi_\alpha^y : [0,\infty) \to [0,\infty)$ with $\Phi_\alpha^y(0) = 0$,
and sequences $(a_r^y)_{r \geq 1}$ of positive numbers with $\lim a_r^y = \infty$ for
every $y \in E$, such that for all $(\alpha^y)_{y \in E} \in A^E$

(a) $$\lim_{r \to \infty} P_{\alpha, r} \; a_r^{y^2} \int_0^t ds \, \alpha^y(s) \frac{\hat{G}_r^y(s)}{G_{\alpha^y}(s)} \; H_r^y(s) 1_{(\tau_r^{*y} > s , \; H_r^y(s) > \epsilon/a_r^y)} = 0$$

$$(y \in E, t \geq 0, \epsilon > 0) ;$$

(b) $\lim\limits_{r\to\infty} P_{\alpha,r} \left(\left| a_r^{y^2} \int_0^t \bar{d}s \; \alpha^y(s) \dfrac{\hat{G}_r^y(s)}{G_\alpha y(s)} \; H_r^y(s) 1_{(\tau_r^{*y} > s)} - \Phi_\alpha^y(t) \right| > \varepsilon \right) = 0$

$$(y \in E, t \geq 0, \varepsilon > 0) \; ;$$

(c) $\lim\limits_{r\to\infty} P_{\alpha,r}(\tau_r^{*y} > t) = 1$ $\hspace{2cm} (y \in E, t \geq 0) ,$

where

$$H_r^y(s) = \frac{\hat{G}_r^y(s)}{G_\alpha y(s)} \; \frac{1}{\hat{Z}_r^y(s)} \; .$$

Then $(a_r^y(\hat{F}_r^y - F_\alpha y))_{y \in E}$ converges in $P_{\alpha,r}$-distribution to the Gauss process with continuous paths, which has independent components, the y'th component having mean 0 and covariance function

(3.8) $\hspace{2cm} v^y(s,t) = \Phi_\alpha^y(s) G_\alpha y(s) G_\alpha y(t) \hspace{2cm} (s \leq t) \; .$

Proof. Proceeding exactly as in the proof of Theorem 2.2, one finds that conditions (a) and (b) are translations of conditions (a) and (b) from Theorem 1.3 applied to the $P_{\alpha,r}$-martingales Q_r^y as represented by (3.4). Therefore, by Theorem 1.3 , $(a_r^y Q_r^y)_{y \in E}$ converges in distribution to the Gauss-Φ_α process with independent increments. Next observe that for any $y \in E$, $t \geq 0$, $\varepsilon > 0$

$$P_{\alpha,r} \left(\sup_{s \leq t} | a_r^y (1 - \frac{\hat{G}_r^y(s)}{G_\alpha y(s)}) - a_r^y Q_r^y(s) | > \varepsilon \right)$$

$$\leq P_{\alpha,r}(\tau_r^{*y} \leq t) \to 0$$

by condition (c), wherefore also $(a_r^y(1 - \frac{\hat{G}_r^y}{G_\alpha y}))_{y \in E}$ converges in distribution to the Gauss-Φ_α process with independent increments. Since

$$a_r^y(\hat{F}_r^y - F_\alpha y) = a_r^y(G_\alpha y - \hat{G}_r^y) = G_\alpha y(a_r^y(1 - \frac{\hat{G}_r^y}{G_\alpha y})) \; ,$$

the assertion of the theorem now follows from the observation that the mapping $f: D[0,\infty)^E \to D[0,\infty)^E$ given by

186

5.3.6

$$(f(w))^y(t) = G_\alpha y(t) w^y(t)$$

is (Skorokhod)-continuous, so that $(a_r^y(\hat{F}_r^y - F_\alpha y))_{y \in E}$ converges in distribution to the process which is the Gauss-Φ_α process with independent increments transformed by f. Since f acts on each component separately, the independence of components is preserved, and it is also clear that the transformed process is Gaussian with mean 0. Finally, the covariance function for the y'th component of the new process is

$$v^y(s,t) = G_\alpha y(s) G_\alpha y(t) v_0^y(s,t) \qquad (s \le t) ,$$

with v_0^y the covariance function for the Gauss-Φ_α process with independent increments. But

$$v_0^y(s,t) = \Phi_\alpha^y(s) \qquad (s \le t) ,$$

and the theorem is proved. ∎

Instead of verifying the somewhat forbidding looking conditions of Theorem 3.7, one may apply the following result.

3.9. <u>Corollary</u>. Suppose that conditions (a) and (b) of Theorem 2.2 hold. Then the conclusion of Theorem 3.7 is valid, with (a_r^y) and Φ_α the same as in Theorem 2.2, provided

(a) $\quad \lim\limits_{r\to\infty} \int_0^t ds\, \alpha^y(s)\, P_{\alpha,r}\, \dfrac{1}{\tilde{Z}_r^y(s)}\, 1_{(\tilde{Z}_r^y(s) > 0)} = 0 \qquad (y \in E, t \ge 0),$

(b) $\quad \lim\limits_{r\to\infty} P_{\alpha,r}(\tau_r^{*y} > t) = 1 \qquad (y \in E, t \ge 0).$

<u>Remark</u>. Condition (a) is the condition for uniform consistency in Proposition 2.25.

<u>Proof</u>. We show that (a) and (b) of Theorem 3.7 are implied by the conditions of the corollary. But fixing $\alpha, r, y, t, \varepsilon$ and omitting subscripts

etc., the expectation in (a) of Theorem 3.7 is, using the monotonicity of $G, \hat{G}$ and the inclusion $(\tau_r^{*y} > s) \subset (Z(s) > 0)$,

$$P_{\alpha, r} \int_0^t ds \; \alpha(s) \left(\frac{\hat{G}(s)}{G(s)}\right)^2 \frac{a^2}{Z(s)} 1_{(\tau^* > s, H(s) > \frac{\varepsilon}{a})}$$

$$\leq P_{\alpha, r} \int_0^t ds \; \alpha(s) \frac{1}{G^2(t)} \frac{a^2}{Z(s)} 1_{(0 < Z(s) < \frac{1}{a} \varepsilon G(t))}$$

which tends to 0 as $r \to \infty$, by (a) of Theorem 2.2.

   To evaluate the probability in (b) of Theorem 3.7, first notice that by Doob's inequality and Lemma 3.3,

$$P_{\alpha, r} \sup_{s \leq t} Q^2(s) \leq 4 \; P_{\alpha, r} \; Q^2(t)$$

$$= 4 \; P_{\alpha, r} <Q>_t$$

$$= 4 \; P_{\alpha, r} \int_0^t ds \; \alpha(s) \left(\frac{\hat{G}(s)}{G(s)}\right)^2 \frac{1}{Z(s)} 1_{(\tau^* > s)}$$

$$\leq \frac{4}{G^2(t)} P_{\alpha, r} \int_0^t ds \; \alpha(s) \frac{1}{Z(s)} 1_{(Z(s) > 0)}$$

which tends to 0 by condition (a) of the corollary. In particular the supremum then tends to 0 in probability, so to verify (b) of Theorem 3.7 it is, also invoking condition (b) of the corollary, enough to show that

$$(3.10) \quad P = P_{\alpha, r} \left(\left| \int_0^t ds \; \alpha(s) \left(\frac{\hat{G}(s)}{G(s)}\right)^2 \frac{a^2}{Z(s)} 1_{(\tau^* > s)} - \Phi^Y(t) \right| > \varepsilon \right. ,$$

$$\left. \sup_{s \leq t} Q^2(s) \leq \delta, \tau^* > t \right) \to 0$$

for a suitable $\delta > 0$. But under the conditions on the supremum and τ^*,

$$I_t = \int_0^t ds \, \alpha(s) \left(\frac{\hat{G}(s)}{G(s)}\right)^2 \frac{a^2}{Z(s)} 1_{(\tau^* > s)}$$

$$= \int_0^t ds \, \alpha(s) \left(1 - \left(1 - \frac{\hat{G}(s)}{G(s)}\right)\right)^2 \frac{a^2}{Z(s)} 1_{(\tau^* > s)}$$

satisfies

$$(1-\sqrt{\delta})^2 \int_0^t ds\, \alpha(s)\, \frac{a^2}{Z(s)}\, 1_{(Z(s)\,>\,0)} \leq I_t \leq \int_0^t ds\, \alpha(s)\, \frac{a^2}{Z(s)}\, 1_{(Z(s)\,>\,0)} \; ,$$

so that for the probability P from (3.10) we get the inequality

$$P \leq P_{\alpha,r}\left(\int_0^t ds\, \alpha(s)\, \frac{a^2}{Z(s)}\, 1_{(Z(s)\,>\,0)} > \Phi^y(t) + \varepsilon\right)$$

$$+ P_{\alpha,r}\left(\int_0^t ds\, \alpha(s)\, \frac{a^2}{Z(s)}\, 1_{(Z(s)\,>\,0)} < \frac{1}{(1-\sqrt{\delta})^2}(\Phi^y(t) - \varepsilon)\right),$$

which by (b) of Theorem 2.2 tends to 0 if $\delta > 0$ is chosen so small
that

$$\frac{1}{(1-\sqrt{\delta})^2}\left(\Phi^y(t) - \varepsilon\right) < \Phi^y(t). \qquad\qquad \Box$$

For Theorem 3.7 and Corollary 3.9 to be of any use, we must estimate the asymptotic covariance function (3.8). Under the conditions
of the theorem, $\hat{G}_r^y$ is consistent for $G_\alpha y$, so it is natural to e-
stimate $\Phi_\alpha^y(s)$ by $\hat{\omega}_r^{y2}$ as defined before Proposition 2.24, and then
estimate $v^y(s,t)$ by

$$\hat{v}^y(s,t) = \hat{\omega}_r^{y2}\, \hat{G}^y(s)\hat{G}^y(t) \qquad\qquad (s \leq t),$$

which will be consistent under the hypothesis of Theorem 3.7 and Proposition 2.24(b).

Remark. Because $G_\alpha y(t) = e^{-\int_0^t \alpha^y}$ one might also estimate $G_\alpha y$ by
$\hat{G}_r^y(t) = e^{-\hat{\beta}_r^y(t)}$. Under the conditions of Theorem 2.2

$$a_r^y(1 - \frac{\hat{G}_r^y}{G_\alpha y}) = a_r^y(1-\exp(-(\hat{\beta}_r^y - \beta_{\alpha,r}^{*y}))) \sim a_r^y(\hat{\beta}_r^y - \beta_{\alpha,r}^{*y})$$

converges in distribution to the Gauss-Φ_α process with independent increments, and it follows that under the conditions of the corollary,
$(a_r^y(\hat{F}_r^y - F_\alpha y))_{y\in E}$ and $(a_r^y(\hat{F}_r^y - F_\alpha y))_{y\in E}$ have the same limit process.

|

For the three examples discussed in the previous section, condition (a) of Corollary 3.9 is satisfied, and since (b) is easily verified, the corollary applies in all three cases. We list the results below.

3.11. Example. For the i.i.d. case of Example 2.6, we get that $\sqrt{r}(\hat{F}_r - F_\mu)$ converges in distribution to the Gaussian mean 0 process with covariance function

$$V(s,t) = \frac{F_\mu(s)}{G_\mu(s)} G_\mu(s) G_\mu(t) = F_\mu(s) G_\mu(t) \qquad (s \leq t),$$

which is the well-known result on the asymptotic behavior of the empirical distribution function, and may be derived directly from a central limit theorem for i.i.d variables. ▮

3.12. Example. For the i.i.d. case with censoring of Example 2.14, we find that $\sqrt{r}(\hat{F}_r - F_\mu)$ converges in distribution to the Gaussian mean 0 process with covariance function

$$V(s,t) = \left(\int_0^s du\, \mu(u)\, \frac{1}{\varphi(u) G_\mu(u)} \right) G_\mu(s) G_\mu(t) \qquad (s \leq t).$$

We have thus described the asymptotic behavior of the Kaplan-Meier estimator. ▮

3.13. Example. For the Markov chain case of Example 2.17, we get that $(\sqrt{r}(\hat{F}_r^{ij} - F_{\mu_{ij}}))_{(i,j) \in E}$ converges in distribution to the Gaussian mean 0 process with independent components and covariance function

$$v^{ij}(s,t) = \left(\int_0^s du\, \frac{\alpha_{ij}(u)}{p_i(u)} \right) G_{\alpha_{ij}}(s) G_{\alpha_{ij}}(t) \qquad (s \leq t)$$

for the (i,j)'th component.

But in this case it is perhaps more interesting to study the asymptotics of $\hat{F}_r^{i\cdot}(t) = 1 - \prod_{s \leq t}(1 - \hat{\mu}_{i\cdot}(s))$, which is the estimator of the

distribution governing the waiting times in state i . The result is
that $(\sqrt{r}(\hat{F}_r^{i\cdot} - F_{\mu_i}))_{i \in S}$ converges in distribution to the Gaussian
mean 0 process with independent components and covariance function

$$v^{i\cdot}(s,t) = \left(\int_0^s du \frac{\mu_i(u)}{p_i(u)}\right) G_{\mu_i}(s) G_{\mu_i}(t) \qquad (s \le t)$$

for the i'th component.   ∎

5.4. Comparison of two intensities.

Suppose given two one-dimensional product processes P_{α^1,r^1} P_{α^2,r^2} with intensities

$$\lambda_t^{1,i} = \alpha^1(t) z_t^{1,i} \qquad (1 \le i \le r^1),$$

$$\lambda_t^{2,j} = \alpha^2(t) z_t^{2,j} \qquad (1 \le j \le r^2).$$

Then the product process $P_{\alpha^1,\alpha^2} = P_{\alpha^1,r^1} \otimes P_{\alpha^2,r^2}$ is a counting process with intensity $(\lambda_t^{1,i}, \lambda_t^{2,j})_{i,j}$. (If $r = r_1 = r_2$ we just have a product of r two-dimensional counting processes).

We shall consider the statistical model obtained when $\alpha^1, \alpha^2 \in A$. Based on observation of P_{α^1,α^2} over an interval $[0, t_0]$, we want to test the hypothesis

(4.1) $\qquad H_0: \qquad \alpha^1(s) = \alpha^2(s) \qquad (0 \le s \le t_0).$

For $k = 1,2$ we shall write

$$\tilde{N}^k = \sum_{i=1}^{r_k} N^{k,i}, \quad \tilde{Z}^k = \sum_{i=1}^{r_k} Z^{k,i},$$

$$\tilde{M}^k(t) = \tilde{N}^k(t) - \int_0^t ds\, \alpha^k(s) \tilde{Z}_s^k\, 1_{(\tilde{Z}_s^k > 0)}$$

and also put $\tilde{N} = \tilde{N}^1 + \tilde{N}^2$, $\tilde{Z} = \tilde{Z}^1 + \tilde{Z}^2$. Notice that $\tilde{M}^1$ and $\tilde{M}^2$ are orthogonal P_{α^1,α^2}-martingales with

$$< \tilde{M}^k >_t = \int_0^t ds\, \alpha^k(s) \tilde{Z}_s^k\, 1_{(\tilde{Z}_s^k > 0)} \qquad (k = 1,2).$$

We know that for $k = 1,2$

$$\beta_{\alpha^k}^{*k}(t) = \int_0^t ds\, \alpha^k(s)\, 1_{(\tilde{Z}_s^k > 0)}$$

is estimated by

$$\hat{\beta}^k(t) = \int_{(0,t]} \tilde{N}^k(ds) \frac{1}{\tilde{Z}^k_{s-}} 1_{(\tilde{Z}^k_{s-} > 0)} .$$

It is also clear that under H_0, with α the common α^1, α^2,

$$\beta^*_\alpha(t) = \int_0^t ds\, \alpha(s)\, 1_{(\tilde{Z}_s > 0)}$$

is estimated by

$$\hat{\beta}(t) = \int_{(0,t]} \tilde{N}(ds) \frac{1}{\tilde{Z}_{s-}} 1_{(\tilde{Z}_{s-} > 0)} .$$

The test of the null-hypothesis (4.1) is based on the following idea: the intensities α^1 and α^2 can only be compared at time-instants s where they are both estimable, i.e. at s such that $\tilde{Z}^1 \tilde{Z}^2(s) > 0$. Now let $K = K_{r^1 r^2}$ be a locally uniformly bounded, predictable process and define

$$T_t = \int_{(0,t]} (\hat{\beta}^1(ds) - \hat{\beta}^2(ds)) K(s) 1_{(\tilde{Z}^1 \tilde{Z}^2(s-) > 0)}$$

$$= \int_{(0,t]} \tilde{N}^1(ds) \frac{K(s)}{\tilde{Z}^1(s-)} 1_{(\tilde{Z}^1 \tilde{Z}^2(s-) > 0)}$$

$$- \int_{(0,t]} \tilde{N}^2(ds) \frac{K(s)}{\tilde{Z}^2(s-)} 1_{(\tilde{Z}^1 \tilde{Z}^2(s-) > 0)} ,$$

so that K weights locally the difference between the estimators for α^1 and α^2. We see that

$$T_t = \int_{(0,t]} \left(\tilde{M}^1(ds) \frac{1}{\tilde{Z}^1(s-)} - \tilde{M}^2(ds) \frac{1}{\tilde{Z}^2(s-)} \right) K(s) 1_{(\tilde{Z}^1 \tilde{Z}^2(s-) > 0)}$$

$$+ \int_0^t ds\, (\alpha^1(s) - \alpha^2(s)) K(s)\, 1_{(\tilde{Z}^1 \tilde{Z}^2(s) > 0)} ,$$

and consequently, under H_0, T is a martingale with

$$< T >_t = \int_0^t ds\, \alpha(s)\, K^2(s) \left(\frac{1}{\tilde{z}^1(s)} + \frac{1}{\tilde{z}^2(s)} \right) 1_{(\tilde{z}^1 \tilde{z}^2(s) > 0)} \; .$$

For the test of H_0, T_{t_0} is used as test-statistic, and the hypothesis is rejected if the observed value of T_{t_0} deviates too much from 0. To carry out the test in practice one needs of course an approximation to the true distribution of T_{t_0} under H_0, and this may be obtained using the limit theorem from Section 5.1. This way one can easily prove the next result, where we assume that $r^1 \to \infty$, $r^2 \to \infty$ in a specific fashion, namely $r^1 = \gamma^1 r$, $r^2 = \gamma^2 r$ where $\gamma^1 \gamma^2 > 0$ and $r \to \infty$.

4.2. **Proposition.** Suppose there exists a sequence (a_r) of positive numbers with $a_r \to \infty$, and a non-decreasing, continuous function $\Phi : [0,\infty) \to [0,\infty)$ with $\Phi(0) = 0$, such that

(a)   For $k = 1,2$ and all $t \geq 0$, $\varepsilon > 0$

$$\lim_{r \to \infty} P_{\alpha^k, \gamma^k_r} \; a_r^2 \int_0^t ds\, \alpha^k(s)\, \frac{1}{\tilde{z}^k(s)}\, 1_{(0 < \tilde{z}^k(s) < \frac{\gamma^k a_r}{\varepsilon})} = 0 \; ;$$

(b)   for all $t \geq 0$, $\varepsilon > 0$ under H_0

$$\lim_{r \to \infty} P_{\alpha\alpha}(| < T >_t - \Phi(t) | > \varepsilon) = 0 \; ,$$

(c)   for every $t \geq 0$ there exists a constant $c_t > 0$ such that

$$\sup_{s \leq t, r} |K_{\gamma^1 r, \gamma^2}(s)| \leq c_t \; .$$

Then, under H_0, T converges in distribution to the Gauss-Φ process with independent increments.

Proof.   Condition (a) just states that condition (a) of Theorem 2.2 holds for both product processes P_{α^1, r^1}, P_{α^2, r^2}. Combining this with the uniform boundedness in (c), one checks that condition (a) of

Theorem 1.3 holds, when applied to the $P_{\alpha\alpha}$-martingale T. █

    If the conditions of the proposition are satisfied, one works with the test-statistic T_{t_0} as if, under H_0, it is normally distributed with mean 0 and variance $\Phi(t_0)$. By the standard methods, the asymptotic variance $\Phi(t)$ is estimated by

$$\hat{\omega}_T^2(t) = \int\limits_{(0,t]} \tilde{N}(ds) \, \frac{K(s-)^2}{\tilde{Z}(s-)} \left(\frac{1}{\tilde{Z}^1(s-)} + \frac{1}{\tilde{Z}^2(s-)} \right) 1_{(\tilde{Z}^1\tilde{Z}^2(s-) > 0)}$$

$$= \int\limits_{(0,t]} \tilde{N}(ds) \, \frac{K(s-)^2}{\tilde{Z}^1\tilde{Z}^2(s-)} \, 1_{(\tilde{Z}^1\tilde{Z}^2(s-) > 0)} \, ,$$

and conditions for this to be consistent may be found as in Section 5.2.

    The choice of the weight process K depends on which alternatives one wishes the test to be powerful against. Thus, if for $k = 1,2$, P_{α^k, r^k} corresponds to observing r^k i.i.d. lifetimes with intensity α^k, different choices of K gives the Wilcoxon test and Savage test respectively for non-parametric comparison of two distributions.

    The results presented here are due to Aalen (1978). The problem of comparing more than two intensities, has recently been solved by Andersen, Borgan, Gill and Keiding (1981).

Notes.

It has been shown that the mean convergence of the sum in condition
(a) of Theorem 1.3 may be replaced by convergence in probability, see
e.g. Shiryayev (1981), Section 6, Corollary 1.

Theorem 2.2 is essentially Theorem 6.4 of Aalen (1978), while
Propositions 2.24 and 2.25 correspond to Propositions 6.5 and 6.3 of
the same paper.

Without altering the assumptions, the conclusion of Proposition
2.24 (b) may be sharpened to an assertion about uniform consistency,
see Exercise 1 below.

Some of the steps in the argument for condition (a) of Theorem 2.2
in Example 2.17, may be omitted: use the lower bound for $\frac{1}{r} S_r^i(s)$ on
p. 5.2.11 directly to get the inequality for $P_{\mu, r}(\exp(-S_r^i(s)))$ on p.
5.2.12.

For applications, it is of course vital that the conditions of
Theorem 5.2.2 can be verified. A different method from the one employed
in the examples in Section 5.2, has been proposed by Aalen and Johansen
(1978), Theorem 4.1. There the critical step consists in showing that
the family

$$\alpha^Y(s) \; \frac{a_r^{y^2}}{\widetilde{Z}_r^y(s)} \; 1\,(\widetilde{Z}_r^y(s) > 0)$$

of random variables obtained for $r \geq 1$, $s \leq t$ be uniformly integrable. Then for instance (a) of Theorem 5.2.2 holds if for each fixed
s, the integrand converges to 0 in probability: the expectation
$E_r(s)$ of the integrand will then converge to 0 because of the uniform integrability when r varies and the operations of integrating
s from 0 to t and letting $r \to \infty$ may be interchanged, since using
the uniform integrability in s also, the convergence of $E_r(s)$ is
dominated for $s \leq t$.

Section 5.3 is an attempt to treat systematically the asymptotics
of a collection of one-dimensional product-limit estimators, but the
results do not cover matrix valued product-limit estimators such as

for instance the estimators for Markov chain transition matrices from Section 4.4. For this particular case, the limiting distributions have been found by Aalen and Johansen (1978).

The asymptotic distribution of the Kaplan-Meier estimator, (Example 5.3.12), was first mentioned by Efron (1967) and a proof was given by Breslow and Crowley (1974).

In the Cox regression model, it is most relevant to discuss the asymptotic properties of the estimator for the regression parameter β, obtained from the partial likelihood (see Section 4.5). For results about this, see Andersen and Gill (1981), Tsiatis (1981).

Some examples of Aalen models not treated in the text, may be found in Exercises 6 and 7 below. For an interesting and difficult example, arising when observing a Markov chain only partially, see Borgan and Ramlau-Hansen (1982).

From the Aalen estimators or product-limit estimators, it is possible to derive asymptotic confidence bands over an interval for the unknown integrated intensity or for the survivor function determined by this. Some references are Gillespie and Fisher (1979), Fleming et al. (1980), Hall and Wellner (1980), Burke et al. (1981), Csörgö and Horváth (1982). One possible approach is discussed in Exercise 8 below.

Since the applicability of the Aalen models rests on the asymptotic theory, it is of interest to know how quickly the limit results apply. For strong approximation theorems providing rates of convergence, see Burke et al. (1981), Csörgö and Horváth (1981), (1982), Földes (1981), Földes and Rejtö (1981).

Some data applications of Aalen models appear in Aalen (1978), Aalen et al. (1980), Andersen, Borgan et al. (1982). See also Aalen (1981).

A famous application of the Cox model appears in Crowley and Hu (1977). Recent examples are Andersen and Rasmussen (1982), Drzewiecki and Andersen (1982). On how to test the appropriateness of the Cox model, see Andersen (1982).

Becker and Hopper (1981) on the infectiousness of the common cold on Tristan da Cunha, is a recent example of the use of martingale methods in applied statistics.

For the statistical analysis of survival data, one major reference is Kalbfleisch and Prentice (1980).

For inference in parametric point process models, relating in particular to time series analysis, see the survey by Brillinger (1978).

Exercises.

1. Show that the conclusion of Proposition 5.2.24 may be sharpened to the following form of uniform consistency: under the conditions of the proposition

$$\lim_{r \to \infty} P_{\alpha,r}(\sup_{s \leq t} |\hat{\omega}_r^{y^2}(s) - \Phi^y(s)| > \varepsilon) = 0$$

for all $y \in E$, $t \geq 0$, $\varepsilon > 0$.

Hints: the basic idea exploits that $\hat{\omega}_r^{y^2}$ and Φ^y are increasing, and that Φ^y is continuous. Fix α, y, let $\varepsilon > 0$, $t > 0$ and choose $0 = s_0 < \cdots < s_n = t$ so that $\Phi(s_k) - \Phi(s_{k-1}) < \frac{\varepsilon}{2}$ for $k = 1, \cdots, n$. Then argue that

$$(\sup_{s \leq t}(\hat{\omega}_r^{y^2}(s) - \Phi^y(s)) > \varepsilon)$$

$$\subset \bigcup_{k=1}^{n} (\hat{\omega}_r^{y^2}(s_k) - \Phi^y(s_{k-1}) > \varepsilon)$$

$$\subset \bigcup_{k=1}^{n} (\hat{\omega}_r^{y^2}(s_k) - \Phi^y(s_k) > \frac{\varepsilon}{2}),$$

and deduce from this that

$$P_{\alpha,r}(\sup_{s \leq t}(\hat{\omega}_r^{y^2}(s) - \Phi^y(s)) > \varepsilon) \to 0.$$

Give a similar argument that

$$P_{\alpha,r}(\inf_{s \leq t}(\hat{\omega}_r^{y^2}(s) - \Phi^y(s)) < -\frac{\varepsilon}{2}) \to 0. \qquad |$$

2. Show that if condition (c) of Theorem 5.3.7 and (5.2.26) hold, then

$$\lim_{r \to \infty} P_{\alpha,r}(\sup_{s \leq t} |\hat{G}_r^y(s) - G_\alpha y(s)| > \varepsilon) = 0$$

for every $y \in E$, $t \geq 0$, $\varepsilon > 0$, i.e. $\hat{G}_r^y$ is a uniformly consistent estimator for $G_\alpha y$.

Hint: show that this amounts to proving

$$P_{\alpha,r}(\sup_{s \leq t} G_{\alpha}y(s) \, |Q_r^y(s)| > \varepsilon) \to 0$$

for all y, t, ε. Argue that this follows from

$$P_{\alpha,r}(\sup_{s \leq t} |Q_r^y(s)| > \varepsilon) \to 0$$

for all y, t, ε, and deduce this last result using Doob's martingale inequality and (5.2.26) as in the proof of Corollary 5.3.9.

∎

3. Verify that the condition stated after Proposition 5.2.24,

$$\lim_{r \to \infty} \sup_{s \leq t} P_{\alpha,r} \frac{a_r^y{}^4}{(\tilde{Z}_r^y(s))^3} 1_{(\tilde{Z}_r^y(s) > 0)} = 0,$$

is satisfied in Examples 5.2.6, 5.2.14, 5.2.17.

∎

4. Show that τ_r^{*y}, as defined in the beginning of Section 5.3, is a stopping time, by showing that

$$(\tau_r^{*y} < t) = \bigcup_{q < t} (\tilde{Z}_r^y(q) < 1),$$

where the union extends over rational q.

∎

5. Show the assertion in Example 5.3.13 about the asymptotic distribution of $(\sqrt{r}(\hat{F}_r^{i \cdot} - F_{\mu_i}))_{i \in S}$.

Hint: reasoning as in Lemma 5.3.3, represent

$$1 - \frac{\hat{G}_r^{i \cdot}(t \wedge \tau_r^{*i})}{G_{\mu_i}(t \wedge \tau_r^{*i})}$$

as a stochastic integral with respect to the martingale $\tilde{M}^{i \cdot} = \sum_{j \neq i} \tilde{M}^{ij}$. Here

$$\tau_r^{*i} = \inf\{t \geq 0: S_r^i(t) = 0\}. \qquad \blacksquare$$

6. Let $X_1, \cdots, X_r$ be i.i.d. lifetimes with intensity $\mu \in A$ and consider the following form of right-censoring: with $X_{(1)} < \cdots < X_{(r)}$ the ordered X_i, only the values of $X_{(1)}, \cdots, X_{(m)}$ are observed, where $m \leq r$ is given. Define

$$K_t = \sum_{j=1}^{m} 1_{(X_{(j)} \leq t)}.$$

Show that the intensity for the CCP generated by K is

$$\lambda_{t-} = \mu(t-)(r-N_{t-})1_{(N_{t-} < m)}.$$

Consider the full Aalen model corresponding to this, when μ varies freely in A, find the Aalen estimator $\hat{\beta}_t$ of

$$\beta_t^* = \int_0^t ds\, \mu(s) 1_{(N_s < m)},$$

and find an unbiased estimator of the mean squared error function

$$\sigma^2(t) = P_\mu(\hat{\beta}_t - \beta_t^*)^2.$$

Suppose now that $r \to \infty$, $m \to \infty$ in such a way that $m/r \to p$ where $0 < p < 1$. Let for $\mu \in A$, π_μ denote the smallest p-fractile for F_μ, i.e.

$$\pi_\mu = \inf\{t \geq 0: F_\mu(t) = p\}.$$

Show that as $r \to \infty$, $\sqrt{r}(\hat{\beta} - \beta^*)$ converges in distribution to the Gauss-Φ process with independent increments, where

$$\Phi(t) = \frac{F_\mu(t \wedge \pi_\mu)}{G_\mu(t \wedge \pi_\mu)}. \qquad \blacksquare$$

7. Consider a Markov process $X = (X_t)_{t \geq 0}$ with state-space $\mathbb{N}_0$ such that, in the notation of Example 2.1.19,

$$\mathbb{P}(T_{n+1} > t | T_1, \cdots, T_n, J_1, \cdots, J_n) = \left(\frac{G(t)}{G(T_n)}\right)^{J_n} \qquad (t \geq T_n) ,$$

$$\mathbb{P}(J_{n+1} = j | T_1, \cdots, T_{n+1}, J_1, \cdots, J_n) = \pi_{j-J_n+1}(T_{n+1}-) \qquad (j \geq J_n-1) .$$

where G is a survivor function and for every $t > 0$, $x \in \mathbb{N}_0$, $\pi_x(t-) \geq 0$ with $\sum_{x \geq 0} \pi_x(t-) = 1$. We shall also assume that $\pi_1(t-) = 0$, so that always $J_{n+1} \neq J_n$.

With this structure X is a Markov <u>branching process</u>, where G is the survivor function for the distribution of the lifetime of an individual born at time 0, and where $\pi_x(t-)$ is the conditional probability that an individual, given that she dies at time t, at that instant gives birth to and is replaced by x newborn individuals.

As in the general Markov case, we shall from now on assume that X_0 is degenerate and equal to r, where $r \geq 1$.

From X we extract a multivariate counting process $K = (K^y)_{y = -1,1,1,\cdots,}$, where for each y and t, K_t^y counts the number of $y+1-$ births on $(0,t]$, i.e.

$$K_t^y = \sum_{s \leq t} 1_{(\Delta X_s = y)} ,$$

so that $X_t = r + \sum_y y \, K_t^y$.

In order to keep K finite-dimensional, we shall assume that the offspring distributions have finite support, i.e. we shall assume that for some $m \in \mathbb{N}$, $\pi_x(t-) = 0$ for $x > m$ and all t.

Suppose now that G has intensity $\mu \in A$ and let the $\pi_x(t-)$ for $x = 0,2,\cdots,m$ be left-continuous in t with right limits.

Show that the CCPE with type-set $E = \{-1, 1, \cdots, m-1\}$, which is generated by K, has intensity $\lambda_- = (\lambda_-^y)_{y \in E}$, where

(1)
$$\lambda_{t-}^y = \mu(t-) \pi_{y+1}(t-) J_{t-}^+ ,$$

writing $J_t = r + \sum_{y \in E} y\, N_t^y$ for the canonical branching process corresponding to N. (Of course $J_t \geq 0$ a.s.).

Show that the basic assumptions from Section 4.1 are satisfied for the intensities specified by (1) with $\alpha^y = \mu \pi_{y+1}$.

Consider the full Aalen model determined by (1). Find for each $y \in E$, the Aalen estimator $\hat{\beta}_t^y$ for

$$\beta_t^{*y} = \int_0^t ds\; \mu(s) \pi_{y+1}(s) 1_{(J_s > 0)} ,$$

and also propose an estimator for $\beta_t^{*\cdot} = \sum_y \beta_t^{*y}$.

Instead of this full Aalen model, we might have considered a product model corresponding to r i.i.d. independent copies of the branching process starting with just 1 individual. The intensity of the i'th sub-counting process is then given by

$$\lambda_{t-}^{i,y} = \mu(t-) \pi_{y+1}(t-) J_{t-}^{i+} ,$$

where $J_t^i = 1 + \sum_y y\, N_t^{i,y}$.

Use Theorem 4.2.3 and the characterization of counting processes through their intensities to show the fundamental branching property: the distribution of the process $(J_t)_{t \geq 0}$ where $J_0 = r$, is the same as the distribution of the process $(J_t^1 + \cdots + J_t^r)_{t \geq 0}$.

Show that the Aalen estimators in the product model and the full model above agree.

The remainder of the exercise is devoted to a study of the asymptotic behavior of the Aalen estimator as $r \to \infty$. For this it is convenient to use the product model rather than the full model (1). Referring to the product model, introduce

$$\tilde{J}_t = \sum_{i=1}^{r} J_t^i ,$$

so that $\tilde{J}$ plays the same part as the original J .

Show that with

$$\xi(t) = P\tilde{J}_t = r \ PJ_t^1$$

(which of course depends on μ and the π_x), the process $(\sqrt{r}(\hat{\beta}_r^y - \beta_r^{*y}))_{y \in E}$ converges in distribution as $r \to \infty$ to the Gauss-Φ process with independent increments, where $\Phi = (\Phi^y)_{y \in E}$ is given by

$$\Phi^y(t) = \int_0^t ds \ \mu(s) \ \pi_{y+1}(s) \ \frac{1}{\xi(s)} .$$

Hint: verify (b) of Theorem 5.2.2 and (5.2.3) adapting the technique from the examples in Section 5.2, and proving and using for instance the bound

$$J_s^{i+} \geq 1_{(\tau_1^i > s)} \geq 1_{(\tau_1^i > t)} ,$$

valid for all $i = 1, \cdots, r$ and all $s \leq t$.

Find an estimator for each $\Phi^y(t)$.

Discuss the asymptotic behavior of the estimator for $\beta^{*\cdot}$.

Following the methods of Section 5.3, find an estimator for $G_\mu(t)$, and derive its asymptotic properties.

Finally, interpret the estimators found above as parameters for a suitable, purely discrete Markov branching process.

For a more general treatment of estimation in Markov branching

processes, allowing arbitrary offspring distributions, see
Harrington and Fleming (1978), Johansen (1981a). |

8.   The purpose of this exercise is to show how to construct asympto-
      tic confidence bands for the estimated integrated intensity in a
      product Aalen model.

To make things simple, we shall assume that $|E| = 1$, i.e. we
have a product of one-dimensional processes.

Let $B = (B_t)_{t \geq 0}$ be a standard one-dimensional Brownian motion
defined on some probability space $(\Omega, A, \mathbb{P})$. Also, let f be a
given non-negative function defined on $[0, \infty)$.

For $t \geq 0, \gamma \geq 0$ introduce

(1) $\psi_f(\gamma, t) = \mathbb{P}(|B_s| \leq \gamma f(s), s \leq t)$.

In principle at least, ψ_f is thus a known function with
$\psi_f(\gamma, 0) = 1$.

Suppose from now on that f is continuous with $f(t) > 0$ for all
$t \geq 0$ (so in particular $f(0) > 0$). Then ψ_f is jointly contin-
ous in γ and t ,   strictly increasing in γ and strictly decreasing in t with

$$\lim_{\gamma \downarrow 0} \psi_f(\gamma, t) = 0 , \lim_{\gamma \uparrow \uparrow \infty} \psi_f(\gamma, t) = 1 .$$

With these assumptions, (1) may be used to construct f-shaped
bands, inside which B will stay on $[0, t]$ with some prescribed
probability 1-p: simply solve the equation $\psi_f(\gamma, t) = 1-p$ for
γ .

Let now $\Phi: [0, \infty) \to [0, \infty)$ be non-decreasing and continuous with
$\Phi(0) = 0$. Then $M_t = B_{\Phi(t)}$ defines a Gauss-Φ process M with

independent increments. Show that

$$\mathbb{P}\left(|M_s| \le \gamma f(\Phi(s)), \ s \le t\right) = \psi_f(\gamma, \Phi(t)).$$

Consider the product Aalen model from Sections 4.2 and 5.2 with $|E| = 1$. Assume that the conditions of Theorem 5.2.2 are satisfied, so that $_aM_{\alpha,r}$ converges in $P_{\alpha,r}$-distribution to M. Assume also that

(2) $$P_{\alpha,r}(\sup_{s \le t} |\hat{\omega}_r^2(s) - \Phi(s)| > \varepsilon) \to 0$$

as is the case if in addition to (a), (b) of Theorem 5.2.2, also (b) of Proposition 5.2.24 is satisfied, cf. Exercise 1 above.

Given f continuous as before, and given $0 < p < 1$, $t > 0$, define for every r, γ_r as the unique solution to

$$\psi_f(\gamma_r, \hat{\omega}_r^2(t)) = 1 - p$$

if $\hat{\omega}_r^2(t) > 0$ and let $\gamma_r = 0$ if $\hat{\omega}_r^2(t) = 0$. Further, let γ denote the solution to

$$\psi_f(\gamma, \Phi(t)) = 1 - p$$

if $\Phi(t) > 0$ and put $\gamma = 0$ otherwise. Note that since Φ depends on the unknown intensity α, so does γ.

The remainder of the exercise is devoted to a proof of the following result: write

(3) $$\pi_{\alpha,r} = P_{\alpha,r}\left(\hat{\beta}_r(s) - \frac{\gamma_r}{a_r} f(\hat{\omega}_r^2(s)) \le \beta^*_{\alpha,r}(s) \le \hat{\beta}_r(s) + \frac{\gamma_r}{a_r} f(\hat{\omega}_r^2(s)), \ s \le t\right)$$

$$= P_{\alpha,r}\left(|_aM_{\alpha,r}(s)| \le \gamma_r f(\hat{\omega}_r^2(s)), \ s \le t\right).$$

Then for all $\alpha \in A$ such that $\Phi(t) > 0$,

(4)
$$\lim_{r \to \infty} \pi_{\alpha,r} = 1 - p .$$

From the convergence in distribution of $a_r M_{\alpha,r}$ to M, it follows that

(5)
$$\lim_{r \to \infty} P_{\alpha,r}(|a_r M_{\alpha,r}(s)| \leq \gamma f(\Phi(s)), \ s \leq t) = \psi_f(\gamma, \Phi(t)) ,$$

a fact that may be assumed in the sequel. (For a formal proof one shows, that with respect to the distribution of M, considered as a probability on the Skorokhod space $D[0,\infty)$, the indicator function on $D[0,\infty)$ determined by the event

$$(|M_s| \leq \gamma f(\Phi(s)), \ s \leq t)$$

is almost surely continuous).

Use the continuity and monotonicity properties of ψ_f to show that

(6)
$$P_{\alpha,r}(|\gamma_r - \gamma| > \varepsilon) \to 0$$

as $r \to \infty$ for every $\varepsilon > 0$.

Fix $\alpha \in A$. Given $\varepsilon > 0$ define

$$\eta(\varepsilon) = \sup_{\substack{x,y \leq \Phi(t)+\varepsilon \\ |x-y| \leq \varepsilon}} |\log f(x) - \log f(y)| ,$$

and use (3), (5) and (6) to show that

$$\lim_{r \to \infty} \sup \pi_{\alpha,r} \leq \psi_f((\gamma+\varepsilon)e^{\eta(\varepsilon)}, t) ,$$

$$\lim_{r \to \infty} \inf \pi_{\alpha,r} \geq \psi_f((\gamma-\varepsilon)e^{-\eta(\varepsilon)}, t) ,$$

the latter if $\varepsilon \leq \gamma$. Let $\varepsilon \downarrow 0$ to obtain (4), using the properties of f.

Thus the left and right sides of the inequalities in the expression defining $\pi_{\alpha,r}$, give asymptotic f-shaped, 1-p level confidence bands on $[0,t]$ for the unknown integrated intensity. If

$$\lim_{r \to +\infty} P_{\alpha,r}(\tilde{Z}_r(s) > 0, \; s \leq t) = 1,$$

one may of course replace $\beta^*_{\alpha,r}(s)$ in (3) by the true integrated
intensity $\int_0^s \alpha$.

We have assumed that $f(0) > 0$, which means that even for $s = 0$
the band has strictly positive width although we know of course
that $\hat{\beta}_r(0) = \beta^*_{\alpha,r}(0) = 0$. But the assumption is critical for
the proof and indeed, if $f(0) = 0$, one has typically that
$\pi_{\alpha,r} = 0$ because $\hat{\beta}_r(s) = \hat{\omega}_r^2(s) = 0$ for small $s > 0$. Note also,
that to obtain $\psi_f(\gamma,t) > 0$ for $\gamma > 0$, $t > 0$ when $f(0) = 0$,
$f(t)$ must rise sharply from 0, more sharply than

$$K\sqrt{2t \; \log \log \frac{1}{t}}$$

for any $K < 1$ as is seen from the law of the iterated logarithm
for Brownian motion.

The same methods used in this exercise may be applied to derive
confidence bands for the $G_\alpha y$ estimated in Section 5.3. ∎

APPENDIX

1. The principle of repeated conditioning.

Let (Ω, A) be a measurable space. A sub σ-algebra $\mathcal{D}$ of A is
separable if there is a countable collection of sets $D_n \in \mathcal{D}$ such that
$\mathcal{D}$ is the smallest σ-algebra containing all D_n.

The atoms of a separable σ-algebra $\mathcal{D}$ are defined to be the non-
empty sets of the form $\underset{n}{\cap} D_n'$ where for every n, $D_n' = D_n$ or $D_n' = D_n^c$.
Clearly the atoms are the smallest non-empty sets in $\mathcal{D}$. Notice that
there may be more than countably many atoms.

By considering the class E of sets which are finite intersections
of some of the D_n, one obtains a determining class, i.e. any finite
positive measure on $(\Omega, \mathcal{D})$ is uniquely determined by its restriction
to E. Of course E is countable.

Let S be a Polish space (i.e. a separable metric space, metris-
able so as to become complete) equipped with its Borel σ-algebra (the
σ-algebra generated by the open sets).

We shall now assume that Ω is a Borel subset of S with A the
σ-algebra of Borel subsets (in S) of Ω.

It is then true that A is separable, and it can be shown that
any separable sub σ-algebra $\mathcal{D}$ of A is saturated, i.e. if $D \in A$
is a possibly uncountable union of atoms for $\mathcal{D}$, then automatically
$D \in \mathcal{D}$. (In terms of an equivalence relation, $\mathcal{D}$ may be described as
follows: $\mathcal{D}$ consists of those A-measurable sets which are unions of
equivalence classes for the equivalence relation $\underset{\mathcal{D}}{\sim}$ on Ω given by
$\omega \underset{\mathcal{D}}{\sim} \omega'$ iff ω and ω' belong to the same $\mathcal{D}$-atom).

Suppose $\mathcal{D}$ is separable. Then if $\mathbb{P}$ is an arbitrary probability
on (Ω, A) there exists a regular proper conditional probability $\mathbb{P}^{\mathcal{D}}$
of $\mathbb{P}$ given $\mathcal{D}$, i.e. $\mathbb{P}^{\mathcal{D}} : \Omega \times A \to [0,1]$ with

(i) $\omega \to \mathbb{P}^{\mathcal{D}}(\omega, A)$ $\mathcal{D}$-measurable for all $A \in A$,

(ii) $A \to \mathbb{P}^{\mathcal{D}} (\omega, A)$ a probability for all $\omega \in \Omega$,

(iii) $\int_{D} \mathbb{P} (d\omega) \, \mathbb{P}^{\mathcal{D}} (\omega, A) = \mathbb{P} (AD)$ for all $A \in A$, $D \in \mathcal{D}$,

(iv) $\mathbb{P}^{\mathcal{D}} (\omega, D_{\omega}) = 1$ for all $\omega \notin N$ where D_{ω} is the $\mathcal{D}$-atom con-
         taining ω ,   and $N \in \mathcal{D}$ satisfies $P(N) = 0$.

Now let $\mathcal{B}, C$ be two separable sub σ-algebras of A . The smallest
σ-algebra $\mathcal{B} \vee C$ containing both of them is then also separable with
atoms each of which is the non-empty intersection of a $\mathcal{B}$-atom with a
C-atom.

Suppose $\mathbb{P}^{\mathcal{B}}$ is a regular, proper conditional probability of $\mathbb{P}$
given $\mathcal{B}$.   Then for $\omega \notin N$ where $N \in \mathcal{B}$ with $\mathbb{P}(N) = 0$, $\mathbb{P}^{\mathcal{B}} (\omega, B_{\omega}) = 1$,
so $\mathbb{P}^{\mathcal{B}} (\omega, \cdot)$ may be viewed as a probability on $(B_{\omega}, B_{\omega} A)$,   and since
B_{ω} is a Borel subset of a Polish space, we know that there exists a
regular, proper conditional probability of $\mathbb{P}^{\mathcal{B}} (\omega, \cdot)$ given $B_{\omega} C$. De-
noting this object by $\mathbb{P}^{\mathcal{B}|C}_{\omega} (\cdot, \cdot)$ it is seen that in particular it has
the following properties: for $\omega \notin N$

(v) $\mathbb{P}^{\mathcal{B}|C}_{\omega} (\omega', A)$ is defined for all $\omega' \in B_{\omega}$, $A \in B_{\omega} A$,

(vi) $\omega' \to \mathbb{P}^{\mathcal{B}|C}_{\omega} (\omega', A)$ is $B_{\omega} C$- measurable for all $A \in B_{\omega} A$,

(vii) $A \to \mathbb{P}^{\mathcal{B}|C}_{\omega} (\omega', A)$ is a probability on B_{ω} for all $\omega' \in B_{\omega}$,

(viii) $\int_{C} \mathbb{P}^{\mathcal{B}} (\omega, d\omega') \, \mathbb{P}^{\mathcal{B}|C}_{\omega} (\omega', A) = \mathbb{P}^{\mathcal{B}} (\omega, AC)$ for all $A \in \mathcal{B}_{\omega} A$, $C \in B_{\omega} C$.

It is convenient to rewrite the last property as

(viii)* $\int_{B_{\omega} C} \mathbb{P}^{\mathcal{B}} (\omega, d\omega') \, \mathbb{P}^{\mathcal{B}|C}_{\omega} (\omega', B_{\omega} A) = \mathbb{P}^{\mathcal{B}} (\omega, B_{\omega} AC)$ for all $A \in A$, $C \in C$.

We are now ready to state the <u>principle of repeated conditioning</u>.

<u>Theorem</u>.   Let $\mathbb{P}$ be a probability on (Ω, A) ,   and let $\mathcal{B}, C$ be sepa-
rable sub σ-algebras of A .   If $\mathbb{P}^{\mathcal{B}}$ and $\mathbb{P}^{\mathcal{B} \vee C}$ are regular, proper
conditional probabilities of $\mathbb{P}$ given $\mathcal{B}$ and $\mathcal{B} \vee C$ respectively,
there exists $N \in \mathcal{B}$ with $\mathbb{P}(N) = 0$ such that

$$\mathbb{P}_\omega^{B|C}(\omega',A) = \mathbb{P}^{B\vee C}(\omega',A) \qquad (\omega' \in B_\omega, \quad A \in B_\omega A)$$

for $\omega \notin N$ defines a regular, proper conditional probability of $\mathbb{P}^B(\omega,\cdot)$ given $B_\omega C$.

Proof. As a function of ω', $\mathbb{P}_\omega^{B|C}(\omega',A)$ is $B \vee C-$, hence $B_\omega C-$ measurable, and it is obviously a probability on B_ω as a function of A, which is concentrated on the $B\vee C$-atom containing ω', which is the same as the $B_\omega C$-atom containing ω'.

The theorem will therefore be proved if we verify (viii)[*]. Since for $\omega' \in B_\omega$, $A \in A$, $\mathbb{P}_\omega^{B|C}(\omega',B_\omega A) = \mathbb{P}_\omega^{B|C}(\omega',A)$ (because $\mathbb{P}^{B\vee C}(\omega',\cdot)$ is concentrated on the $B\vee C$-atom containing ω' which is a subset of $B_{\omega'} = B_\omega$), and since $\mathbb{P}^B(\omega,B_\omega) = 1$ for ω outside a P-null set in B, we must show that there exists $N \in B$ with $\mathbb{P}(N) = 0$ such that for $\omega \notin N$

(1) $$\int_C \mathbb{P}^B(\omega,d\omega') \mathbb{P}_\omega^{B|C}(\omega',A) = \mathbb{P}^B(\omega,AC)$$

simultaneously for all $A \in A$, $C \in C$.

Fix $A \in A$, $C \in C$. Then

$$\int_C \mathbb{P}^B(\omega,d\omega') \mathbb{P}_\omega^{B|C}(\omega',A) = \int_C \mathbb{P}^B(\omega,d\omega') \mathbb{P}^{B\vee C}(\omega',A)$$

is the conditional expectation with respect to $\mathbb{P}$ of $1_C \mathbb{P}^{B\vee C}(\cdot,A)$ given B. Consequently for any $B \in B$

$$\int_B \mathbb{P}(d\omega) \int_C \mathbb{P}^B(\omega,d\omega') \mathbb{P}_\omega^{B|C}(\omega',A) = \int_B \mathbb{P}(d\omega) 1_C(\omega) \mathbb{P}^{B\vee C}(\omega,A)$$

$$= \int_{BC} \mathbb{P}(d\omega) \mathbb{P}^{B\vee C}(\omega,A)$$

$$= \mathbb{P}(ABC)$$

$$= \int_B \mathbb{P}(d\omega) \mathbb{P}^B(\omega,AC).$$

Therefore, for every $A \in A$, $C \in C$ there is a $\mathbb{P}$-null set $N_{AC} \in B$ such that (1) holds for all $\omega \notin N_{AC}$. Letting A vary in a countable determining class A^* for A and C vary in a countable determining class C^* for C, we can find $N \in B$ with $\mathbb{P}(N) = 0$ such that for $\omega \notin N$, (1) holds simultaneously for all $A \in A^*$, $C \in C^*$. But for A, respectively C, fixed, both sides of (1) is a finite measure as a function of C, respectively A, and so (1) holds for $\omega \notin N$, simultaneously for all $A \in A$, $C \in C$. ▌

Replacing the σ-algebras A, B by random variables X, X', the theorem may somewhat loosely be stated as follows: conditioning first on $X = x$, and then, inside this first conditioning, conditioning also on $X' = x'$, is the same as conditioning on $(X, X') = (x, x')$.

References on saturation and proper regular conditional probabilities are Blackwell (1956) and Parthasarathy (1967), (in particular Section V.8).

2.   Weak convergence.

Let S be a metric space and let $\mathcal{S}$ denote the Borel σ-algebra on S . Write C(S) for the space of bounded, continuous functions from S to $\mathbb{R}$.

Let P_n, P be probabilities on (S,$\mathcal{S}$) .

1. <u>Definition</u>. $P_n \Rightarrow P$ (P_n converges weakly to P) if $P_n(f) \underset{n \to \infty}{\to} P(f)$ for all $f \in C(S)$. |

This definition leads to the concept of convergence in distribution: let $(\Omega_n, A_n, \mathbb{P}_n)$ be a sequence of probability spaces, let for each n , X_n be a S-valued random variable defined on (Ω_n, A_n) . Also let $(\Omega, A, \mathbb{P})$ be a probability space, X a S-valued random variable defined on (Ω, A) .

2. <u>Definition</u>. $X_n \overset{\mathcal{D}}{\to} X$ (X_n converges in distribution to X) if $\mathbb{P}_n X_n^{-1} \Rightarrow \mathbb{P} X^{-1}$. |

Of course $\mathbb{P}_n X_n^{-1}$ is the distribution of X_n , i.e. the probability on S induced by X_n from the probability $\mathbb{P}_n$ on Ω_n .

Definition 2 requires that

$$\mathbb{P}_n f(X_n) \underset{n \to \infty}{\to} \mathbb{P} f(X) \qquad\qquad (f \in C(S)).$$

We shall discuss convergence in distribution of stochastic processes (real-valued, time-axis $[0,\infty)$) having right-continuous, left-limit paths or just continuous paths.

Denote by $D[0,\infty)$ the space of paths w: $[0,\infty) \to \mathbb{R}$ which are right-continuous with left-limits, by $C[0,\infty)$ the space of continuous paths w: $[0,\infty) \to \mathbb{R}$. Also for t > 0 , define $D[0,t]$ as the space of paths w: $[0,t] \to \mathbb{R}$ which are right-continuous with left-limits.

Similarly, introduce $C[0,t]$.

Topologies on $C[0,t]$ and $D[0,t]$. For $w_1, w_2 \in C[0,t]$ define

$$d(w_1, w_2) = \sup_{s \leq t} |w_1(s) - w_2(s)| .$$

Then d is a metric for the uniform topology on $C[0,t]$. With this metric, $C[0,t]$ is a complete, separable metric space.

We shall now discuss the Skorokhod topology on $D[0,t]$. The topology should have the following property: if $(w_n), w$ is a sequence of paths with only one jump from 0 to 1 at $(t_n), t$, then $w_n \to w$ if $t_n \to t$. This means that we cannot use the uniform topology on $D[0,t]$, but must allow perturbations of time. Denote by Ψ_t the space of functions $\psi : [0,t] \to [0,t]$ which are continuous, strictly increasing with $\psi(0) = 0$, $\psi(t) = t$. Then define

$$d(w_1, w_2) = \inf\{\varepsilon > 0 : \exists \psi \in \Psi_t \text{ with } \sup_{s \leq t} |\psi(s) - s| \leq \varepsilon ,$$

$$\sup_{s \leq t} |w_1(s) - w_2(\psi(s))| \leq \varepsilon\} .$$

This d is a metric, and the topology on $D[0,t]$ generated by d is the Skorokhod $D[0,t]$-topology. A metric d_0 , equivalent to d , can be found, such that with d_0 , $D[0,t]$ is a complete separable metric space. $\quad$ |

Convergence in $D[0,t]$. For a sequence $(w_n), w \in D[0,t]$, we have $w_n \to w$ iff there exists $\psi_n \in \Psi_t$ such that $\psi_n(s) \to \psi(s)$ and $w_n(\psi_n(s)) \to w(s)$ uniformly in $s \in [0,t]$. Furthermore, if $w_n \to w$, then $w_n(s) \to w(s)$ for s a continuity point of w . $\quad$ |

Topologies on $C[0,\infty)$ and $D[0,\infty)$. For a path w in $C[0,\infty)$ or $D[0,\infty)$, write $r_t w$ for the restriction of w to $[0,t]$.

On $C[0,\infty)$ we shall use the topology of uniform convergence on compact sets. Thus

$$w_n \to w \text{ in } C[0,\infty) \text{ iff } r_t w_n \to r_t w \text{ in } C[0,t] \text{ for all } t \geq 0 .$$

To get a topology on $D[0,\infty)$, introduce Ψ as the space of continuous and strictly increasing functions $\psi: [0,\infty) \to [0,\infty)$ with $\psi(0) = 0$, $\lim_{t\to\infty} \psi(t) = \infty$. We shall then demand that $w_n \to w$ in $D[0,\infty)$ iff there is a sequence (ψ_n) of functions in Ψ such that $\psi_n(s) \to s$ uniformly in $s \geq 0$ and $w_n(\psi_n(s)) \to w(s)$ uniformly in s on any compact subset of $[0,\infty)$.

This convergence gives the <u>Skorokhod $D[0,\infty)$ - topology</u>. As in the case of $D[0,t]$: if $w_n \to w$, then $w_n(s) \to w(s)$ for s a continuity point of w . With a suitable metric for the topology, $D[0,\infty)$ becomes a complete, separable metric space. |

<u>Weak convergence of probabilities on $C[0,\infty)$ and $D[0,\infty)$</u> . Having defined the topologies on $C[0,\infty)$ it also makes sense to talk about weak convergence of probabilities on the two spaces. Thus if (P_n) , P are probabilities on $C[0,\infty)$ or $D[0,\infty)$

(3) $$P_n \to P \text{ if } P_n(f) \to P(f)$$

for all bounded, continuous $f: C[0,\infty) \to \mathbb{R}$, respectively $f: D[0,\infty) \to \mathbb{R}$.

For $k \in \mathbb{N}$ and $0 \leq t_1 < \cdots < t_k$ denote by $\pi_{t_1 \cdots t_k}$ the projection $\pi_{t_1 \cdots t_k}(w) = (w(t_1), \cdots, w(t_k))$ from $C[0,\infty)$ or $D[0,\infty)$ to $\mathbb{R}^k$. Then $\pi_{t_1 \cdots t_k}$ is continuous on $C[0,\infty)$, so taking $f = g \circ \pi_{t_1 \cdots t_k}$ in (3), where $g \in C(\mathbb{R}^k)$, it follows that if (P_n) , P are probabilities on $C[0,\infty)$ with $P_n \to P$, then the finite-dimensional distributions of P_n converge weakly to those of P .

The projection $\pi_{t_1 \cdots t_k}$ is not continuous on $D[0,\infty)$: $\pi_{t_1 \cdots t_k}$ is continuous at $w \in D[0,\infty)$ iff w is continuous at $t_1, \cdots, t_k$. So the result above about finite-dimensional distributions on $C[0,\infty)$ must be rephrased in the case of $D[0,\infty)$.

Suppose (P_n), P are probabilities on $D[0,\infty)$ with $P_n \to P$. Define

$$T_P = \{t \geq 0: P\text{-almost all } w \text{ are continuous at } t\}.$$

Then $P_n \pi_{t_1 \cdots t_k}^{-1} \to P \pi_{t_1 \cdots t_k}^{-1}$ if $t_1, \cdots, t_k \in T_P$. In particular, if P is concentrated on $C[0,\infty)$: $P(C[0,\infty)) = 1$, then the finite-dimensional distributions of P_n converge weakly to those of P.

In the applications we make, the limiting probability will always be concentrated on $C[0,\infty)$. So consider a sequence (P_n) of probabilities on $D[0,\infty)$ and a probability P on $C[0,\infty)$. Then, in order that $P_n \to P$ it is necessary and sufficient that the finite-dimensional distributions of P_n converge weakly to those of P _and_ that the sequence (P_n) be _relatively compact_, i.e. each subsequence $(P_{n'})$ of (P_n) contains a further subsequence $(P_{n''})$ which is weakly convergent.

Thus, to show that $P_n \to P$ one must prove that the finite-dimensional distributions converge, and that (P_n) is relatively compact. The latter is accomplished by showing that (P_n) is _tight_, i.e. for every $\epsilon > 0$ there exists $K \subset D[0,\infty)$ compact such that $P_n K > 1-\epsilon$ for all n, and then using

4. _Theorem_ (Prokorov). A family of probability measures on a complete, separable metric space is relatively compact if and only if it is tight. ▮

Criteria for tightness are given in Billingsley (1968) for the spaces $C[0,1]$ and $D[0,1]$, and for $C[0,\infty)$ and $D[0,\infty)$ in Lindvall (1973). The following theorem is from that paper. We assume that (P_n), P are probabilities on $D[0,\infty)$.

5. <u>Theorem</u>. $P_n \to P$ if and only if $P_n r_t^{-1} \to P r_t^{-1}$ for all $t \in T_P$. $\blacksquare$

Of course, if P is a probability on $C[0,\infty)$, the condition becomes $P_n r_t^{-1} \to P r_t^{-1}$ for all $t \geq 0$. $\blacksquare$

<u>Convergence in distribution of stochastic processes</u>. Suppose that $(\Omega_n, A_n, \mathbb{P}_n)$, $(\Omega, A, \mathbb{P})$ are probability spaces, and let (X_n), X denote $D[0,\infty)$-valued random variables, X_n defined on Ω_n, X on Ω. In accordance with Definition 2, X_n converges in distribution to X, $X_n \overset{D}{\to} X$, if $\mathbb{P}_n X_n^{-1} \to \mathbb{P} X^{-1}$.

But a $D[0,\infty)$-valued random variable is nothing but a stochastic process with paths that are right-continuous with left-limits. So one can now talk about convergence in distribution of a sequence of such processes. $\blacksquare$

For further information about weak convergence, see the references Billingsley (1968) and Lindvall (1973) already mentioned above.

References

Aalen, O.O. (1975). Statistical inference for a family of counting processes. Ph.D. dissertation, University of California, Berkeley.

Aalen, O.O. (1976). Nonparametric inference in connection with multiple
decrement models. Scand. J. Statist. $\underline{3}$, 15-27.

Aalen, O.O. (1977). Weak convergence of stochastic integrals related to
counting processes. Z. Warsch. Verw. Gebiete $\underline{38}$, 261-277. Correct-
ion: ibid. $\underline{48}$ (1979), 347.

Aalen, O.O. (1978). Nonparametric inference for a family of counting
processes. Ann. Statist. $\underline{6}$, 701-726.

Aalen, O.O. (1980). A model for nonparametric regression analysis of
counting processes. Proceedings, Sixth International Conference
on Mathematical Statistics and Probability Theory, Wisla (Poland)
1978. Klonecki, W., Kozek, A., Rosinski J. (eds). Lecture Notes
in Statistics 2, pp. 1-25, Springer, New York.

Aalen, O.O. (1981). Practical applications of the nonparametric theory
for counting processes. Technical report, University of Tromsø.

Aalen, O.O., Borgan, Ø., Keiding, N., Thormann, J. (1980). Interaction
between life history events. Nonparametric analysis for prospective
and retrospective data in the presence of censoring. Scand. J.
Statist. $\underline{7}$, 161-171.

Aalen, O.O., Johansen, S. (1978). An empirical transition matrix for
non-homogeneous Markov chains based on censored observations.
Scand. J. Statist. $\underline{5}$, 141-150.

Andersen, P.K. (1982). Testing goodness- of-fit of Cox's regression
and life model. Biometrics (to appear).

Andersen, P.K., Borgan, Ø., Gill, R.D., Keiding, N. (1982). Linear nonparametric tests for comparison of counting processes, with applications to censored survival data. Internat. Statist. Rev. (to
appear).

Andersen, P.K., Gill, R.D. (1981). Cox's regression model for counting
processes: a large sample study. Research report 81/6. Statistical
Research Unit, Danish Medical and Social Science Research Councils.

Andersen, P.K., Rasmussen, N.K. (1982). Admission to psychiatric hospitals among women giving birth and women having induced abortion. Research report. Statistical Research Unit, Danish Medical and Social Science Research Councils.

Bailey, K.R. (1979). The general maximum likelihood approach to the Cox regression model. Ph.D. dissertation, University of Chicago, Chicago, Illinois.

Becker, N., Hopper, J. (1981). The infectiousness of a disease in a community of households. (Submitted to Biometrika).

Billingsley, P. (1968). Convergence of probability measures. Wiley, New York.

Blackwell, D. (1956). On a class of probability spaces. Proc. Third Berkeley Symp. on Math. Statist. and Probab. Vol. II, pp 1-6. J. Neyman (ed). University of California Press, Berkeley.

Boel, R., Varaiya, P., Wong, E. (1975). Martingales on jump processes I: Representation results. II: Applications. SIAM J. Control. $\underline{13}$, 999-1021 and 1022-1061.

Borgan, Ø., Ramlau-Hansen, H. (1982). Estimation of intensities via cumulative incidence rates. A counting process approach. Technical report. Laboratory of Actuarial Mathematics, University of Copenhagen. (In preparation).

Brémaud, P. (1972). A martingale approach to point processes. Electronics Research Laboratory, Memo #M-345, University of California, Berkeley.

Brémaud, P., Jacod, J. (1977). Processus ponctuels et martingales: résultats récents sur le modélisation et filtrage. Adv. in Appl. Probab. $\underline{9}$, 362-416.

Breslow, N.E. (1975). Analysis of survival data under the proportional hazards model. Internat. Statist. Rev. $\underline{43}$. 45-57.

Breslow, N., Crowley, J. (1974). A large sample study of the life table and product limit estimates under random censorship. Ann. Statist. $\underline{2}$, 437-453.

Brillinger, D. (1978). Comparative aspects of the study of ordinary time series and of point processes. Developments in statistics,

Vol. 1, pp. 33-133. P. Krishnaiah (ed). Academic Press, New York.

Burke, M.D., Csörgö, S., Horváth, C. (1981). Strong approximation of some biometric estimates under random censorship. Z. Wahrsch. Verw. Gebiete 56, 87-112.

Chou Ching-Sung, Meyer, P.-A. (1975). Sur la representation des martingales commes intégrales stochastique dans les processus ponctuels. Séminaire de Probabilités IX. P.A. Meyer (ed). Lecture Notes in Mathematics, Vol. 465, pp. 226-236, Springer, Berlin.

Cox, D.R. (1972). Regression models and lifetables. J. Roy. Statist. Soc. Ser. B 34, 187-220. (With discussion).

Cox, D.R. (1975). Partial likelihood. Biometrika 62, 269-276.

Crowley, J., Hu, M. (1977). Covariance analysis of heart transplant data. J. Amer. Statist. Assoc. 72, 27-36.

Csörgö, S., Horváth, L. (1981). On the Koziol-Green model for random censorship. Biometrika 68, 391-401.

Csörgö, S., Horváth, L. (1982). On cumulative hazard processes under random censorship from the right. Scand. J. Statist. 9, 13-21.

Davis, M.H.A. (1976). The representation of martingales of jump processes. SIAM J. Control 14, 623-638.

Dellacherie, C. (1980). Un survol de la théorie de l'intégrale stochastique. Stochastic Process. Appl. 10 , 115-144. (Also in: Measure theory, Oberwolfach 1979, Proceedings. D. Kölzow (ed.) Lecture Notes in Mathematics, Vol. 794, pp. 365-395. Springer, Berlin (1980)).

Dellacherie, C., Meyer, P.-A. (1975). Probabilités et potentiel. Chapitres I à IV. Hermann, Paris. (English translation: Probabilities and potential. Hermann, Paris, North-Holland, Amsterdam (1978)).

Dellacherie, C., Meyer, P.-A. (1980). Probabilités et potentiel. Chapitres V à VIII. Theorie des martingales. Hermann, Paris.

Drzewiecki, K.T., Andersen, P.K. (1982). Survival with malignant melanoma. Regression analysis of prognostic factors. Cancer (to appear).

Efron, B. (1967). The two sample problem with censored data. Proc. Fifth Berkeley Symp. on Math. Statist. and Probab. Vol. IV, pp.

831-853. L. le Cam and J. Neyman (eds). University of California Press, Berkeley.

Fleming, T.R. (1978a). Nonparametric estimation for nonhomogeneous Markov processes in the problem of competing risks. Ann. Statist. 6, 1057-1070.

Fleming, T.R. (1978b). Asymptotic distribution results in competing risks estimation. Ann. Statist. 6, 1071-1079.

Fleming, T.R., Harrington, D.P. (1981). A class of hypothesis tests for one and two sample censored survival data. Comm. Statist. A-Theory Methods. 10, 763-794.

Fleming, T.R., O'Fallon, J.R., O'Brien, P.C., Harrington, D.P. (1980). Modified Kolmogorov-Smirnov test procedures with applications to arbitrarily right censored data. Biometrics 36, 607-625.

Földes, A. (1981). Strong uniform consistency of the product limit estimator under variable censoring. Z. Wahrsch. Verw. Gebiete 58, 95-107.

Földes, A., Rejtö, L. (1981). A LIL type result for the product limit estimator. Z. Wahrsch. Verw. Gebiete 56, 75-86.

Gill, R.D. (1980a). Censoring and stochastic integrals. Mathematical Centre Tracts 124, Mathematisch Centrum, Amsterdam.

Gill, R.D. (1980b). Nonparametric estimation based on censored observations of a Markov renewal process. Z. Wahrsch. Verw. Gebiete, 53, 97-116.

Gill, R.D. (1981). Testing with replacement and the product limit estimator. Ann. Statist. 9, 853-860.

Gillespie, M.J., Fisher, L. (1979). Confidence bands for the Kaplan-Meier survival curve estimate. Ann. Statist. 7, 920-924.

Hall, W.J., Wellner, J.A. (1980). Confidence bands for a survival curve from censored data. Biometrika 67, 133-143.

Harrington, D.P., Fleming, T.R. (1978). Estimation for branching processes with varying and random environment. Math. Biosci. 39, 255-271.

Jacobsen, M. (1972). A characterization of minimal Markov jump processes. Z. Wahrsch. Verw. Gebiete, 23, 32-46.

Jacobsen, M. (1982). Maximum-likelihood estimation in the multiplicative intensity model. Institute of Mathematical Statistics, University of Copenhagen. (In preparation).

Jacod, J. (1975). Multivariate point processes: predictable projection,
Radon-Nikodym derivatives, representation of martingales. Z. Wahrsch.
Verw. Gebiete, 31, 235-253.

Johansen, S. (1981a). The statistical analysis of a Markov branching
process. Preprint 5, Institute of mathematical statistics, University of Copenhagen. (Submitted to Z. Wahrsch. Verw. Gebiete).

Johansen, S. (1981b). An extension of Cox's regression model. Preprint
11, Institute of mathematical statistics, University of Copenhagen.
(Submitted to Internat. Statist. Rev.).

Kalbfleisch, J.D., Prentice, R.L. (1980). The statistical analysis of
failure time data. Wiley, New York.

Kaplan, E.L., Meier, P. (1958). Nonparametric estimation from incomplete observations. J. Amer. Statist. Assoc. 53, 457-481.

Kiefer, J., Wolfowitz, J. (1956). Consistency of the maximum likelihood
estimator in the presence of infinitely many incidence parameters.
Ann. Math. Statist. 27, 887-906.

Koziol, J.A., Green, S.B. (1976). A Cramèr-von Mises statistic for
randomly censored data. Biometrika 63, 465-474.

Lindvall, T. (1973). Weak convergence in the function space $D[0,\infty)$.
J. Appl. Probab. 10, 109-121.

Liptser R.S., Shiryayev, A.N. (1977-78). Statistics of random processes. Vol. I-II. Springer, Berlin.

Nelson, W. (1969). Hazard plotting for incomplete failure data. J.
Qual. Tech. 1, 27-52.

Oakes, D. (1981). Survival times: aspects of partial likelihood. Internat. Statist. Rev. 49, 235-264. (With discussion).

Parthasarathy, K.R. (1967). Probability measures on metric spaces.
Academic Press, New York.

Rebolledo, R. (1978). Sur les applications de la théorie des martin-
    gales à l'étude statistique d'une famille de processus ponctuels.
    Proceedings, Journées de statistique des processus stochastique,
    Grenoble 1977. D. Dacunha-Castelle, B.v. Cutsem (eds). Lecture
    Notes in Mathematics, 636, pp. 27-70. Springer, New York.

Rebolledo, R. (1980). Central limit theorems for local martingales. Z.
    Wahrsch. Verw. Gebiete 51, 269-286.

Shiryayev, A.N. (1981). Martingales: recent developments and applica-
    tions. Internat. Statist. Rev. 49, 199-233.

Tsiatis, A.A. (1981). The asymptotic distribution of the efficient
    scores test for the proportional hazards model calculated over
    time. Biometrika 68, 311-315.

Williams, D. (ed). (1981). Stochastic integrals. Proceedings, LMS Dur-
    ham Symposium, 1980. Lecture Notes in Mathematics, 851. Springer,
    Berlin.

SUBJECT INDEX

Aalen estimator   4.3.1

Aalen model, basic assumptions   4.1.1
    full   4.1.3
    product   4.2.1

absorption, for a distribution   1.1.1
    for a process   1.2.1

absorption probability   1.1.1

accumulated intensity function   2.5.2

accumulated intensity process   2.5.3

adapted process   3.1.1

atom of σ-algebra   1.2.3, A1

branching process   5.E.4

canonical counting process   1.2.4
    generated   1.2.5
    with type-set E   2.1.2

censoring   1.2.7

compensator   3.1.6

competing risks   4.1.6

confidence band   5.E.7

counting process   1.2.1
    absorption   1.2.1
    canonical   1.2.4
    canonical with type-set E   2.1.2
    full path-space   1.2.2
    of class $\mathcal{D}^E$ 2.5.1
    of class H 1.4.1
    of class H^E 2.2.1
    one-dimensional   1.2.1
    path-space with type-set E   2.1.1
    product of   2.3.1
    stable   1.2.1
    stable canonical   1.2.4
    stable path-space   1.2.3
    with finite expectations locally 1.5.1, 2.2.7

with type-set E	2.1.1
Cox regression model	4.5.1
baseline hazard	4.5.2
$\mathcal{D}^E$, class of counting processes	2.5.1
density	1.1.1
smooth	1.1.2
determining class	A1
distribution, purely discrete	2.5.1
distribution function	1.1.1
Doob-Meyer decomposition	3.1.5
dual predictable projection	3.1.6
evanescent process	3.1.3
example, censored survival times	1.2.7, 2.1.3, 4.1.5, 4.2.2, 4.3.6, 4.6.6, 5.2.7, 5.3.9,
i.i.d. lifetimes	1.2.6, 2.1.2, 2.1.9, 4.1.4, 4.2.2, 4.2.6, 4.3.5, 4.6.4, 5.2.4, 5.2.15, 5.3.9
Markov chains	1.3.2, 1.3.3, 2.1.5, 2.1.8, 4.1.6, 4.4.1, 5.2.10, 5.3.9
exponential law	1.1.2, 2.5.1
filtration	1.2.1
self-exciting	1.2.2
finite expectation locally	1.5.1, 2.2.7
full Aalen model	4.1.3
Gauss-Φ process with independent increments	5.1.2
H, class of counting processes	1.4.1
H^E, class of counting processes	2.2.1
hazard function	1.1.2
increasing process	3.1.2
finite	3.1.2
integrable	3.1.2
locally integrable	3.1.2

indistinguishable processes   1.4.3, 3.1.3

innovation theorem   1.4.12

integrated intensity   1.4.8

intensity function   1.1.2
   accumulated   2.5.2

intensity process   1.4.3, 2.2.3, 2.5.3
   accumulated   2.5.3
   integrated   1.4.8

Kaplan-Meier estimator   1.2.7

left-censoring   4.E.2

likelihood function   1.6.1

locally square integrable martingale 3.1.5

locally uniformly bounded process   3.2.2

local martingale   3.1.5

Markov process   1.2.5
   stationary transitions   1.2.6

martingale   1.5.1, 3.1.4
   local   3.1.5
   locally square integrable   3.1.5

measurable process   3.1.1

natural increasing process   3.1.6

Nelson estimator   4.3.5

optional sampling theorem   3.1.4

orthogonal martingales   2.2.9, 3.1.10

partial likelihood   4.5.3

path-space, for counting process   2.1.1
   with type-set E
   full counting process   1.2.2
   stable counting process   1.2.3

Poisson process   1.2.5
   with type-set E   2.1.2

population at risk   1.2.7

predictable process   2.5.4, 3.1.3

predictable stopping time   3.2.9

pre-τ algebra   1.2.10

principle of repeated conditioning   A2

product Aalen model   4.2.1

product-limit estimator   1.2.8, 4.3.6

products of counting processes   2.3.1

Prohorov's theorem   A8

random time   1.2.9

regular proper conditional probability   A1

relatively compact sequence of probabilities   A8

right-censoring   4.E.1

risk, population at   1.2.7

saturated σ-algebra   A1

separable σ-algebra   A1

Skorokhod topology   A6, A7

statistical model   1.6.1

stochastic integral, definition   3.2.1

stopping time   1.2.9

submartingale   1.5.1, 3.1.4

survivor function   1.1.1

termination point   1.1.1

tight sequence of probabilities   A8

transition intensities for Markov chains   4.1.7

type-set   2.1.1

uniform integrability   3.1.4

uniformly bounded process   3.2.2

Lecture Notes in Statistics

Vol. 1: R. A. Fisher: An Appreciation. Edited by S. E. Fienberg and D. V. Hinkley. xi, 208 pages, 1980.

Vol. 2: Mathematical Statistics and Probability Theory. Proceedings 1978. Edited by W. Klonecki, A. Kozek, and J. Rosiński. xxiv, 373 pages, 1980.

Vol. 3: B. D. Spencer, Benefit-Cost Analysis of Data Used to Allocate Funds. viii, 296 pages, 1980.

Vol. 4: E. A. van Doorn, Stochastic Monotonicity and Queueing Applications of Birth-Death Processes. vi, 118 pages, 1981.

Vol. 5: T. Rolski, Stationary Random Processes Associated with Point Processes. vi, 139 pages, 1981.

Vol. 6: S. S. Gupta and D.-Y. Huang, Multiple Statistical Decision Theory: Recent Developments. viii, 104 pages, 1981.

Vol. 7: M. Akahira and K. Takeuchi, Asymptotic Efficiency of Statistical Estimators. viii, 242 pages, 1981.

Vol. 8: The First Pannonian Symposium on Mathematical Statistics. Edited by P. Révész, L. Schmetterer, and V. M. Zolotarev. vi, 308 pages, 1981.

Vol. 9: B. Jørgensen, Statistical Properties of the Generalized Inverse Gaussian Distribution. vi, 188 pages, 1981.

Vol. 10: A. A. McIntosh, Fitting Linear Models: An Application of Conjugate Gradient Algorithms. vi, 200 pages, 1982.

Vol. 11: D. F. Nicholls and B. G. Quinn, Random Coefficient Autoregressive Models: An Introduction. v, 154 pages, 1982.

Vol. 12: M. Jacobson, Statistical Analysis of Counting Processes. vii, 226 pages, 1982.

Vol. 13: J. Pfanzagl (with the assistance of W. Wefelmeyer), Contributions to a General Asymptotic Statistical Theory. vii, 315 pages, 1982.

Vol. 14: GLIM 82: Proceedings of the International Conference on Generalised Linear Models. Edited by R. Gilchrist. v, 188 pages, 1982.